18194 days life of a pigman

by

Michael James

Preface

From humble beginnings with no farming background. He leaves school to complete an apprenticeship scheme. After managing various units he ends his working life managing a state of the art unit pioneering freedom farrowing. Follow the trials and tribulations, the mishaps, the highlights and the lowlights through his 50 year career in pig farming.

You will cringe, you will laugh, and you will learn a bit along the way about how so many things have improved over the years in the way we look after our pigs.

It's an eye opener for some and a trip down memory lane for others as you follow him through his working, and home life.

Thank you to my daughter, Zoe, for the illustrations.

Part one. The first 11588 days
The end before the beginning

Ten months to 'R' day!

Retirement age! State pension due!

Retirement had not crossed his mind until the two emails popped up in quick succession.

The first one was from the DWP.

"You are due to claim your pension soon, fill in the form on our website", or words to that effect.

The second email was from his works pension provider saying more or less the same thing......Quite a wakeup call to say the least.

How could this have suddenly come around, in his head he was still only forty five! However, every so often his body and aching joints would remind him that forty five was long gone. And for the past few years, as each winter approached, he had been telling himself that he couldn't do the job much longer.

In fact, he had been heard to say, years ago, that he didn't want to be in this line of work after reaching Forty.

Twenty six years on and, yes, still in the same line of work. So much for that idea. Maybe retirement was the way out, a new beginning. Find a part time job, something totally different. A new challenge, yes that would work.

And so, he decided to set a date, later than his official retirement date. Eight months later to be precise.

But now a time to look back at his working life and where it all began, what seemed like a lifetime ago. And on the day he planned to retire it would be eighteen thousand one hundred and ninety four days since he left home to start his fifty year adventure.

Current music trend.... Ed Sheeran.

Fashion......................Drainpipe jeans

The beginning

Once upon a time in an ordinary village, nestled on the north bank of the Humber estuary, lived an ordinary family on an ordinary council house estate. Neither rich nor poor. Neither successful nor unsuccessful. Just surviving happily like everybody else.

The father would cycle off to work each morning leaving at 6 a.m. The seven mile journey taking more out of him day by day. Against all the odds he worked hard and long hours at the specialist steel tube engineering firm. Twenty five long years. And his work ethic was to be inherited by his children.

Polio had struck him and his brother when they were just five years old leaving them both needing a leg calliper on the affected leg. A condition that never got better, only worse.

The mother stayed at home doing what mothers did back then. She had been a seamstress many years before. Now with four children and a husband to look after her working days were long gone.

At the beginning of this tale the first born, a boy, is fifteen years old. His younger brother is thirteen years old. The youngest of the three boys, is ten years old. And the fourth child, a girl, is just four years old. Whether the mother and father meant to stop at three mouths to feed we will never know.

As the years went by and the children grew up, the long idyllic school summer holidays grew shorter and inevitably would end soon. The endless hours of playing cricket on the field down near the foreshore and the rather risky raft building in the natural lake that had formed in the disused quarry, now known as 'Little Switzerland'

Alas, growing up was not negotiable.

And so, as the years drifted by, and the children lost the innocence of youth they began to fly the nest one by one.

The eldest went to college to train as a teacher, a career he was to follow through to his retirement.

Next to fly was the middle brother. Away to university to study a subject none of us understand to this day, however, he also made a lifetime career of teaching, becoming a professor and emigrating to Australia.

Now, to the youngest of the boys. It is fair to say this one was always a bit of a rebel. Not in a malicious or violent way but he just didn't want to conform and was once branded as 'subversive' by Mr Woodward, his chemistry teacher at secondary school. When he looked up the definition in a dictionary, he apparently was a 'leader on of others'. Does that translate as 'management potential'?

Although by no means unintelligent, he was somewhat overshadowed by his elder brothers. Possibly fanning the flames of the rebel in him. This made him less inclined to try hard at school. With a little more effort, he could have excelled in all subjects but distinctly average in most subjects he decided to be. However, when it came to sport and the manual aspects of any subjects, he would throw himself in wholeheartedly. Happier with a spanner or a rugby ball in his hand rather than a pen or pencil.

Two defining moments were to follow, the memories of which would stay with him to this day.

Which came first the writer is not sure, but it is immaterial really. Let us focus first on the collection of his exam results. GCE, s and CSE, s in those days.

As instructed, he walked up to school along with his fellow pupils to collect his results in person. As he approached the main entrance of the school building, he was greeted by his biology teacher, Mr Henderson, a teacher who he had always liked and always got along with.

An opinion that was soon to change!

"Ah......Rhodes.... I have no idea why you have bothered coming for your results, a total waste of time."

Mr Henderson was now no longer his favourite teacher.

Any ideas of staying on at school to progress his education disappeared that instant without even seeing any results, good or bad. As it happened his results were pretty dismal, apart from metalwork, woodwork and English Language!

Defining moment number two. The career advisor's interview. A lady, who appeared too young to be advising on any child's future.

The conversation went something like this,

"So, Michael, what would you like to do when you leave school?"

"I want to be a farmer."

"Ah, why is that?"

"That's what I want to do."

"Does your family farm?"

"No."

"Do you have any connections with farming?"

"My uncle has a sow in his garden."

Rather bemused the advisor tried a new tack.

"Wouldn't you rather go and work in the shipyard or the local supermarket as a trainee manager like so many of your friends?"

This rather dictatorial approach to advising where Michael's future should lay poked a stick at the rebel in him

"No, I am going into farming."

"Well, I have no information to help you with that, I am afraid."

And with that the interview was over. The second defining moment!

Current music trend......Black Sabbath

Fashion....................... Flares and Platform shoes

The Placement and leaving home

And so, readers, Michael's future was set, and I will take over the tale from here. As you may have guessed the writer of this tale is Michael, the youngest of the three boys.

I needed to find a way into farming. No internet searching back then of course, in fact we didn't even have a telephone at home. I found some information at the local library about a three year apprenticeship that the Agricultural Training Board ran. It consisted of two years practical experience working on a farm, followed by one year at Agricultural college studying all aspects of the industry from accounts to how to milk a cow. Upon completion the reward was a National Certificate in Agriculture.

So, after further investigation I enrolled for the course. Resulting from the initial one day initiation meeting with the ATB tutors I was given the opportunity to start work on a mixed arable and livestock farm in a village called Kilham in the heart of the Yorkshire Wolds.

Everything seemed to be moving along quickly now and it felt as though I finally had some direction.

I left home in the August. My few belongings packed into a holdall. ....

A short interlude

Almost four months into my retirement. I thought I best put pen to paper again.

Not 'writer's block', just a series of events, including retirement of course, moving house and a knee replacement, (not mine fortunately).

So back to my tale.

Current music trend......Van Morrison

Fashion...................... Flowery shirts

Work

Time plays tricks with your memory, but I am pretty sure it was the August of 1972 when I left home, just sixteen, but almost seventeen. An epic adventure you may think. At the time it did feel like it. However, the place I was to live and work for the next two years was only thirty two miles from home.

I must admit I do not recall any long goodbyes from my parents. What I do remember is being collected to start my new life, by Toby, the manager of the pig unit that I was to start work on. He arrived in my future boss's Volkswagen Beetle, an import from South Africa where he used to live, and he had used the car in many rallies apparently. It turned out my future boss, Mr David Lazenby, was a bit of a petrol head and loved speed. This, I was to experience firsthand later when he bought one of the first rotary engine Mazdas. The RX3 coupe. But more of that later......

I digress ....

For the next two years my home was to be a two berth, (I use that term lightly), caravan parked at the bottom of the stackyard on the main farm which was on one of the two narrow lanes that led out of the village. West End was where the main farm was. This lane merged on the outskirts of the village with Back Lane, where the pig unit was. If you refer to Google maps you will get a clearer picture of where everything was, although the main farm is now a small housing estate.

My caravan was directly behind the farm foreman's house which sat at the entrance to the farm.

The main farm was devoted to rearing calves for barley beef. Feed for them was grown on the three hundred acres of arable land, growing wheat, barley and oats, all situated close to the farm.

Three 'old boys' looked after the arable and beef side. Not a derogatory term as all three were well into their senior years. They had worked on the farm for many years before it was taken over by David.

George was the foreman. Bernard was head tractor driver and Tom fitted in wherever he was needed. They would tell tales of when all the work on the land was done by horses. So, they were long hard days as the horses had to be looked after before the work began, and after the day was over. Longer days and harder work with little pay.

Three young girls looked after the calves along with David's wife, Cybil. Two lived locally while the third lived in another caravan further up the stackyard. The girls were on a similar scheme to me and would go on to Agricultural college.

The pig unit was situated across a small field opposite the main farm on Back Lane. A rough track ran through the field and connected the two lanes and also provided access to the two main calf rearing buildings which were halfway across the field.

The pig unit consisted of just two hundred breeding sows rearing the progeny to bacon.

And so, my life in agriculture begins and what follows over the ensuing years may or may not be of interest to you dear readers, but if you manage a chuckle here and there or can relate to any particular part of my story then I will have achieved my aim.

Firstly, I would like to describe how I lived for the first two years, bearing in mind I am sixteen, just left school, led a fairly sheltered life and travelled no further than to Skipsea for our annual family holidays. My intention is to hopefully help you to build a picture in your mind from the way my life progressed.

As I have mentioned earlier my humble abode was a two berth caravan. You entered into a small kitchen area which had a small sink and a calor gas hob and oven, neither of which I ever used.

In the 'living area' there were two bench seats doubling as beds. Heating was a coal stove. In winter this packed out a tremendous amount of heat while lit but unfortunately it did not

9

stay in all night, subsequently halfway through the night all the heat was lost, and the caravan became cold and damp. Many a winter's night I would sleep in a sleeping bag with an ex-RAF Greatcoat over me.

One meal a day was provided by the foreman's wife. George and I would go in for dinner around midday. Stevie, as she was known, was a classic farmer's wife, the type that people from towns imagine all farmer's wives are like. She loved cooking and was very good at it. However, the portions were enormous. Three courses at midday were gratefully received but getting back to work after that was a real struggle.

I must have bathed around at their house too or maybe I saved it for my fortnightly visits back home.

Breakfast and tea I might add was 'self-catering'. I lived mainly on milk, Mr Kipling apple pies, chocolate biscuits and condensed milk.

The farm was set on the outskirts of the village. In the village there were two public houses, two shops, a garage and a Church. The Bay Horse pub was set on the corner of a T junction just down from the family run garage and opposite the small Mace shop. It was a Cameron's pub and served a really nice pint of Cameron's Strongarm. Like so many village pubs back then it was a drinker's pub, surviving on liquid sales.

Further down the village and handily situated opposite the Church was the other pub, The Star. This was a 'Free House', a favourite with the older generation in the village. A perfect pint of hand pulled cask Guiness could be quaffed here.

Friday night at The Star was Domino night and I was encouraged to join in when I was there. But I was no match for the 'old boys'

Tom from the farm used to be a regular at The Star on a Friday and Saturday night and would easily manage eight pints of the Irish nectar each night, sitting in the small smoke filled snug, most of the smoke emanated from the pub landlord who seemed to chain smoke filter less Craven A cigarettes. Tom was as thin as a rake but fit as a butcher's dog. He was my favourite of the three that worked on the farm. I must have spent a fair bit of time between the two pubs, not that I was a heavy drinker.

10

The two small shops were a Mace and a Spar. Both were just a short walk from my caravan, but I used the Mace most. Both were very popular shops as not everyone had transport to get into Driffield, the nearest town.

The milk was in glass pint bottles. My guilty secret is for some reason I never returned the bottles. I kept putting them in the cupboard under the sink in my caravan. Finally came a time to clear them out. Sorry but I dumped them in a hedge back up the road away from the farm. Once again sorry to whoever found them.

As I have explained my caravan was at the bottom of the stackyard set to the right of the entrance to the farm. On the left as you entered the farm there were three open straw based yards which housed the calves from the point that they came off milk feeding. Directly opposite my caravan was a small concrete block-built hut where the arable boys would gather at the start of the working day to await David's instructions for the day ahead.

He lived across the road from the farm in a large bungalow.He would come marching into the stackyard bang on 7.25 am., flanked  by his two faithful, yet rather intimidating, Alsatians, Rimsky and Raja. Rimsky was placid by comparison to Raja. Many a time I had Raja snapping at my heels as I exited my caravan. This would set Rimsky off too. It was quite a spine tingling experience. David found that quite amusing.

To the right of my caravan there were two low roofed buildings which formed part of the accommodation for the newly purchased calves which were bought in from a farm down in Devon. They arrived at about four weeks old weighing around fifty kilos. They were kept in individual pens made up of what I would call five bar wooden sheep gates. They were housed like this for several weeks and fed purely on milk substitute. They were bedded up on straw. They had to be trained to suckle, then fed the milk from a bucket, and were looked after by the three young girls. The milk substitute, in a powder form, was called 'Denkavit' and was mixed in a machine then carried to the calves in the individual ten litre blue buckets. There were two more similar buildings in the field across the road from the main farm. These were new and purpose built.

The stackyard on the main farm was just rough crushed limestone and on a fairly steep gradient. (Bare that in mind for later along with the aforementioned wooden calf pen structures).

Halfway up the stackyard on the right was the old granary store, a traditional build with stone steps up to the first-floor storage area. I remember carrying the fifty kilo bags of molasses up those steps. This was used in the feed mix for the barley beef cattle.

At the top of the stackyard was the mill and corn store. To the left of the mill a track led round past a loading dock and sorting area for weighing the barley beef. Just beyond that were the straw based yards holding the beef cattle. These were all under the same roof as the mill and corn store and could be accessed from the mill via an elevated inspection walkway.

Just along West End from the main farm was another series of buildings belonging to the farm. One old building was used to house surplus pigs. Always a bit of a mess. Also, here there was another granary type store similar to the one on the main farm with the stone steps. Just up from the yard from the granary was a large Dutch barn where some of the small straw bales were stacked. Although the majority of the bales were stacked outside in the fields.

All the fields belonging to the farm were fairly close by. All had names. At the top of the hill on the main road out of the village was the largest field, 'Kesters'. This was a large triangular shaped field. Four more fields were at the end of West End on another road leading out of the village. On the left there were three big fields joining each other. A lot of thought had been put into naming these fields, they were, 'Near field', 'Middle field' and 'Far field'. The fourth field, 'Bottom field', a smaller triangular shaped field, was set between the two roads that led out of the village.

The last and largest of the fields, 'Long Flats', was on Back Lane and ran up from where the pig unit was. Almost a mile long rising up like a classic Wolds farm field of clay and chalk.

It's funny, as I am describing the farm, I seem to have a tale to tell about each area, mainly mishaps. All will be revealed later.

And so, to my first year of full-time work to be spent on the pig unit. I had worked during my summer holidays while at school at a Rose growing business, so I was no stranger to work.

Toby, the unit manager, lived in a newly built three bedroom bungalow to the right of the entrance to the unit. A second identical bungalow was to be built shortly after I started working there. This was to be used by my successor and his family.

The whole unit was straw based entirely, even the farrowing houses.

The breeding sows were all tethered. A rudimentary but common practice in those days. The tethers were a simple construction of mild steel rod sleeved with hard plastic tubing. A straight length of rod at each side, one side bent up at the bottom to form a hook. The other side was bent round to form a loop at the bottom. The top of each side rod was also bent to form a loop and secured to the third piece of rod, looped at each end and slightly curved, which rested on the top of the sow's necks. A chain held the tether in place secured to the bottom loop, and on the sow stall or crate. A second chain went from the loop to the hook on the other side to secure the tether around the sow's necks. The chain under the sow's necks was the only way the tether could be adjusted to cater for different size sows. A crude but cheap and common system. Problems did occur mainly as the sows' conditions changed after having the piglets removed at weaning. They soon put on weight once they were served and pregnant again. And so, it was essential to check the tethers regularly. Quite often a sow would be found with the flesh on her neck having grown through the links of the bottom chain. Another problem that occurred occasionally was the hook on the bottom of the tether would catch on the loop on the crate or stall rendering the sow immobile, and if not found soon enough they could easily be strangled to death.

Unfortunately, this system of housing sows was common practice until the late nineties when both stalls and tethers were banned. Subsequently pig producers had to re-think and re-design all of the sow accommodation, although sows in the farrowing houses could still be kept in crates but not tethered. As you can imagine this cost thousands of pounds and caused many producers to go out of business.

I will try and describe how this farm worked as well as I can remember. It will probably shock some people and stir memories for others, however, it was how pigs were kept back then when I started out.

The dry sow accommodation consisted of two long buildings with two rows of tubular steel crates roughly the length of the sows and just wide enough to allow the sow to lay on its side, albeit, tight up against the bars of the crate. All the sows were tethered. The layout, if you were to view the building from one end, was a feed passage, a row of crates, a muck passage, a row of crates and a feed passage. Both buildings had solid concrete floors. For feeding there was a glazed tile trough which ran the full length of each row in front of the sows. They were fed individually from a feed barrow. Water was given twice a day being flushed through one central pipe from a header tank above each row.

At the back of the sows the concrete extended beyond the end of the crate far enough for the sow to lay relatively comfortably, then there was a small step down into the muck passage. Each time the sows were mucked out they were given a small amount of straw behind them. In essence they spent most of their lives on bare concrete.

We used a hand operated Honda engined scraper to push the muck out, this doubled as a rotavator with other attachments.

The boars were housed in pens at one end of both sow houses. All sows were served naturally, no artificial insemination.

Farrowing was on a weekly basis as was weaning and serving.

There were four farrowing houses with four crates either side with a central passage. These buildings were a prefabricated concrete construction with an asbestos roof. I think they had been bungalows in a previous life as post war accommodation in Hull. The farrowing crates were a simple tubular steel construction with a long U-shaped hinged gate at the back. The sows were tethered. The creep area for the piglets was open and to one side of the crate at the head end. Heat for them was just an infra-red lamp hung over that area. Straw was used for bedding in here also. These were mucked out twice a week by hand into a wheelbarrow.

Weaning, that's removing the piglets from their mothers, was done when the piglets were eight weeks old. By modern standards the piglets would be enormous by this age, but they were not as it happens. Partly down to poor knowledge in general of the sow's dietary needs. The sow milk quality and quantity diminished as the piglets got older. Also sow condition suffered which had a knock-on effect on subsequent production. However, this was the preferred method back then. I don't recall the sows being fed on any kind of feed curve, whereby you increase the amount they are fed as the piglets get older.

At weaning the sows would move back to the dry sow accommodation and put into the stalls near the boar pens. The piglets were walked around via a 'run' to the weaner accommodation.

This was a long, low roofed, building raised up from ground level. The whole building was asbestos in wooden frames. There were covered pens either side of a central passageway, with hinged lids. The pens were about eight feet by four feet and about three feet high. At the rear of the pen was a small 'pop hole 'to allow the piglets access to the outside muck area and drinking water nipples. The muck area was expanded steel, (diamond type), raised up above a slightly sloped piece of concrete. This muck area was not covered. The theory was pigs muck in the coolest place. However, quite often they would muck inside. Mucking these out was done by scraping the sloping concrete off with a garden hoe and then pushing it away with the Honda and round to the muck heap which was at the end of the fattening houses. I do not remember there being any feed hoppers in this building, so we were feeding on the floor of the sleeping area.

From the weaner accommodation the pigs were moved at around thirty kilos to the fattening sheds. There were two of these sheds. A fairly standard design. Each building had a central inspection passageway with sleeping area pens either side and a narrow muck passage.

These were the only buildings with an automatic feeding system. It was an E.B. equipment system with adjustable dump hoppers above each pen which were filled by a centreless auger which ran around the whole building. A continuous wire ran around the building attached to each dump hopper. This was pulled by a pulley and motor at the end of each building which lifted the outer part of the dump hopper at set times. The food dropped directly onto the sleeping area floor.

No straw was used in the sleeping area, only in the muck passage, which was scraped out twice a week with the Honda.

As the pigs approached bacon weight, they were weighed individually in the muck passage. If they were in the correct weight range, they were marked with a thick red paint applied on their backs with a piece of wood. Before going off to the abattoir they were slap marked on their shoulders with our farm's identification number.

Bulk feed bins were situated between the buildings holding sow rations, weaner rations and fattener rations.

All the buildings were in line which meant a long passage ran from the sow house to the fattening sheds. A fence made of corrugated tin sheets kept the pigs in when being moved. At the very end was a loading dock which was used for incoming gilts and bacon pigs leaving the unit.

Apart from the Honda scraper used for mucking out there was an old Massey Ferguson 135 on the pig unit, used just for heaping the muck heap up. It was pretty awful to drive. The tires were bald, and the brakes were almost non-existent. It had a fore end loader with basic controls on the quadrant to the right of the seat. This had two levers to lower or raise the loader and the three-point linkage. To tip the muck fork you had to pull another lever which was attached by a wire to the fork which was supposed to be spring loaded to enable it to lock back in position once tipped. This didn't always work so you would have to drop it to the ground to force the locking pin back in. There was no counterbalance weight on the three-point linkage at the back which made the steering very heavy.

I hope that you now have some sort of image building up in your mind as I am now about to attempt to relate to you the first of several mishaps I had over the next two years. Oh yes, please remember I had just left school, was not a farmer's son and had no machinery experience.

This first one was by far the worst. How I kept my job after it I have no idea.

On this particular afternoon I was tasked with heaping up the muck at the end of the fattening houses, which remember, were adjacent to the weaner accommodation, the footprint of which extended beyond the length of the fattening houses.

So, away I go, steady away. Start the 'Fergy' up. Fore end loader down, push in, lift up, push in and tip. Easy. At this point I must point out how slippery concrete can be with years of pig effluent standing on it.

I was doing a pretty good job for a novice, until I got to the end near the weaner accommodation. I needed to drive into the muck heap at a slight angle to create space as this was a path we used regularly. So, I drove in, lifted up, pushed in and tipped then pulled back to go again. Foot on the brakes, nothing happened, still rolling backwards I stood up and jumped on the brake pedal. With that the rear wheels locked but I was now sliding backwards on the slippery concrete, not rolling! Panic ensued as I inched closer, at what was a pretty slow speed, to the end pens of the weaner building. Why didn't I steer? I hear you ask. Well, I did but the front wheels slid too.

With a sickening crunch I hit the building, demolishing the asbestos framework on two pens. The little 'Fergy' coughed and stalled. At this point Toby appeared. He was remarkably calm about what I had done.

"Oh dear, you better go over to the farm and tell David what you have done".

Now for some reason, rather than walk over to the farm I jumped back on the 'Fergy' and trundled over on that. Out onto the road, across the field, passing the calf buildings. Back onto West end and into the farm. One of the calf girls was there looking surprised to see me.

"Do you know where David is?"

"Yes, he is up in the stackyard weighing cattle".

Now, what I should have done was parked the 'Fergy' near my caravan and walked up the stackyard. I chose to drive up. Parking directly outside the corn drier doors. Carefully applying the hand brake and lowering the fore end loader and left it in gear, triple security, I was on a fairly steep slope!

I jumped off the 'Fergy' and went round the corner to find David hurling abuse at a beast that was refusing to go in the weigh crush.

"Yes Mike? What can I do for you?" he snarled.

I explained what had happened. He also was actually quite calm about it, considering his usual fiery temper. There were a few expletives.

"Just get back to work and I will send someone over to have a look."

Well, that wasn't too bad. I walked back round to the 'Fergy' and jumped back on feeling quite relieved at the outcome.

Now, you know when something starts going wrong and you get in a bit of a panic? No? Well listen carefully, and remember I am a novice on a tractor.

To start the tractor up the high and low ratio gear lever had to be in neutral. So, foot on the clutch I pulled back on the lever and engaged neutral, with my foot on the brake of course. I started her up. Released the handbrake and lifted my extra security, the forend loader. The 'Fergy' started rolling backwards, slowly at first, my foot was still hard down on the brake pedal, the handbrake had no effect as I pulled up on it. The forend loader must have been all that was holding it. I started rolling faster. I tried forcing it into gear, but now it was rolling the lever wouldn't move. I tried dropping the forend loader but the gradient on the stackyard was too great for that to hold the 'Fergy'. I was now heading directly to the farm entrance, backwards! I managed to turn the steering wheel and was now heading towards the old granary steps. Well at least they would stop me.

For the second time in an hour, I came to a crunching halt. Unfortunately, just outside the old granary was where the calf girls stacked the wooden calf pen partitions ready for pressure washing.

Not only had I destroyed several of these partitions but also the long lengths of wood that braced them together in situ. I jumped off the 'Fergy' to inspect the devastation.

What a mess! Maybe David would go easy on me, after all accidents do happen. Absolutely no chance. As I gingerly approached where he was still wrestling with his reluctant beasts, I could almost sense he somehow knew what had happened. The air went blue before I even spoke. In many words that I had never heard before he suggested I get out of his sight and get back to the pig unit. I never did have a very smooth relationship with David for a while after that, the first of many mishap induced encounters. The remainder of my year on the pig unit, however, went fairly smoothly as I remember.

I had been earning my first proper weekly wage. £13.50. Presented in cash every Friday in a small brown envelope. With part of the first week's wage, I bought myself a split cane fishing rod. Later I treated my mum and dad to a pressure cooker and a set of pans. I think my dad was only earning twice as much as me at this point. I felt good being able to treat them.

And so, as the first year of my three year course came to an end I moved over from the pigs to the arable and beef side of the business.

This change came in the August, so I was thrown straight into harvest time. Another steep learning curve ahead.

Anyone who was involved in the arable side of agriculture back then may be familiar with the chemical 'Propcorn'. This was sprayed onto the harvested corn as it made its way from the corn trailer to the corn storage bins. Apparently, this process

19

speeded up the drying of the corn. It was extremely toxic and to my knowledge is no longer used.

On one occasion, mid harvest, the elevator that lifted the corn from the pit it had been tipped into, and then onto the conveyor that in turn took the corn to the storage bins, blocked up. The elevator was basically a wide rubber belt with metal cups riveted onto it. And ran vertically on a pulley system.

It was at the bottom of the elevator where the Propcorn spray applicator was. There must not have been a 'fail safe' on the applicator as it had kept spraying even though the elevator had blocked and stopped. George, the farm foreman, had spotted the blockage but not before a puddle of Propcorn had formed in the bottom of the elevator shaft.

The only way to clear the blockage was to actually go down a vertical ladder to the bottom of the narrow elevator shaft and shovel the corn out into buckets tied on a rope so they could be hoisted up. David and George were there, and I was asked to help. It was at this time I think I went up in David's estimation.

I went down the pit first. There was hardly any room to work. The fumes were horrendous. PPE was unheard of. No gloves. No masks. After thirty, maybe forty seconds and two or three buckets cleared, I was feeling lightheaded.

"Right Mike, get back up here now", David shouted, concerned for my safety. I came up gasping for air. Next George went down, but not for long. Then David, then me again. We finally got the blockage cleared. I am pretty sure David stopped using Propcorn that day.

Harvest was a busy but very enjoyable time. Long hours but great fun. Bernard, the second oldest farm hand, drove the combine. A Claas Matador, green, no cab, with a twelve-foot cutting bar. Yes, only twelve feet. George looked after the mill and corn drier. Tom and I, (the Guinness drinker), drove the corn carts. Tom had a Zetor 6011 with, I recall, a stiff column gear change lever. I had the old Fordson Supermajor, the one with the lever you had to push down to start it. It had a homemade plywood cab with faded yellowed Perspex side windows. Both the corn trailers were Weeks' three-ton single axle type. The tailgates were opened and closed by hand. Which reminds me of another little mishap. One which I will list at the end of this chapter, just so I don't forget any.

An old tradition that David and his wife Cybil kept up during harvest was afternoon 'luence'. At around three in the afternoon Cybil would come to the field we were working, everything would stop. We would sit down together and drink the tea she brought in the white enamel jug and eat whatever she had made for us. Sometimes massive teacakes and cheese, or fruit cake and cheese or a generous slice of homemade apple pie. Also, when we finished late in the evening and arrived back at the farm, David would appear with a crate of pint bottles of Courage Pale Ale. These traditions are lost now or are not cost effective. However, it was small things like that that made you feel appreciated.

When harvest was over it was time to prepare the land for the next crop. I was allowed to go muck spreading. Now you may have seen the modern-day muck spreading machines on the road carrying tons of muck, with rotors driven by the tractor PTO shaft, with hydraulic brakes etc. Dismiss that image. My machine was no longer than a family estate car. There were two rotors at the back and the rubber belt that formed the bed of the machine was moved by the wheels as were the rotors. To enable this to happen the tyres on the machine were on the wrong way round to create grip, in theory. All the muck to be spread was already in heaps in each field. I had a small two-wheel drive Ford 3000 tractor to pull the muck spreader and to load up I had my 'favourite', Massey Fergusson, tractor. Now, if you didn't load the machine up carefully it would just bung up. So, you had to layer it from back to front so it would drag itself through the rotors. If it bunged up it was a matter of climbing in and forking it out.

Another job for the arable men, (and boy), was mucking out the calf sheds. All but one could be done with a tractor. The one to the right of my caravan had to be mucked out by hand. The method for this was, two or three of us with muck forks and a couple of wheelbarrows. A small two-wheel muck trailer was parked outside with a wooden plank propped up against the back. The wheelbarrows full of muck were then pushed up the plank. Back breaking and very tiring. However, this was another job that carried a bonus 'thank you' again in the form of a bottle of beer at the end of the day.

In winter jobs on the arable side eased off. But one job I do remember was hedge slashing. This was all done by hand with long handled blade. We would cut the hedges and heap the cuttings up in the field then set fire to. A midwinter job, so I was hot near the fire then freezing cold when I moved away. I think my hatred of the cold stemmed from then.

For the whole of the second year, as part of the course I was on, I attended 'Day Release' which was run by the Agricultural College that I would move onto for the third year of my apprenticeship. The day release was run in the nearest town, Driffield, six miles away.

David usually gave me a lift as I had no transport at this point. This was my first experience of his 'need for speed'. His Mazda RX3 was a real 'pocket rocket'. We touched ninety miles per hour on a road which to anyone else would have been a fifty miles per hour maximum. A true white-knuckle ride. I am not sure how much any of us at 'Day Release' learnt. What I do remember is the cafe we used to call into for our lunch. They made a lovely chips, beans and cheese and onion flan. We also, somehow managed to squeeze in a visit to a pub in the town quite often.

Another experience involving David was when I was asked to go 'bush beating' for him and some of his friends. I went along, it gave me a day off work at least. It was late November, I think. It was very wet underfoot in the fields. I ended up carrying about ten kilos of mud on my boots, not my idea of fun. At lunch time the 'plebs', sorry the 'bush beaters', sat in one corner of a barn while the 'guns' sipped from their hip flasks and ate their cucumber sandwiches. I guess some people love bush beating. I found it very demeaning.

Despite all my accidents and incidents, David was a very generous person not only because of the harvest treats and beer 'thank yous', but to celebrate my eighteenth birthday he and his wife took us all down to The Star for drinks and snacks.

Before I close this first part of my life as a pig farmer, I will briefly turn back to the other mishaps I mentioned earlier. That is apart from the demolition hour that you are already still pondering over.

So, here we go, hopefully in chronological order......

1. A very close shave on the old Massey Furguson as I was exiting the yard at West End where the extra pig yards were. I blamed the brakes again as I was inches away from impaling a fuel delivery tanker as he passed by.

2. Whilst corn carting in harvest with the old Fordson Super major I drew up alongside Bernard on the combine harvester and carefully kept up with him so he could tip and cut at the same time. Bernard started tipping but when he looked to check all was going ok he was shouting and gesticulating. I turned to look at the trailer and realised I had not locked the tailgate shut. Much corn was lost.

3. Just a minor one this one, well minor in so far as I didn't break anything. It might even have been the same day as the open tailgate. I arrived back at the farm with my load of corn, slowly entering the stackyard. David was there. He came over, shouting and swearing at me. (Mind you he did use to shout and swear a lot at Cybil, but always ended his cursing with the word 'darling'. It seemed I was not going fast enough. In his temper he more or less dragged me off the tractor, jumped on, and took over. Leaving me standing there. I think he had been having a bad day.

4. Another corn carting incident. If you recall the field called 'Kesters' at the top of the steep hill going out of the village? I was on the Fordson again with the three-ton corn trailer full

23

coming back from 'Kesters'. With far too much weight behind me for this poor old tractor I was struggling to slow down, in fact the weight of the trailer was speeding me up as I rapidly and was in danger of jack knifing and overtaking me as I approached the junction to West End. Fortunately, by virtually jumping on the brakes I managed to slow down and stop just in time.

5. During harvest again. I was let loose carting straw with the Zetor 6011 pulling a small single axle trailer. Fully loaded I pulled onto the grass field which separated Back Lane from West End. I went for my lunch. When I came back, I jumped on the tractor and pulled forward. There was a hump from the grass onto the track. Now, in hindsight I should have driven up the hump at a right angle with both the tractor and the trailer. I didn't. The tractor negotiated the hump ok but as the near side trailer wheel hit the hump the offside wheel was still on the grass. All was not well, I turned round to see the trailer and its load tipping over sideways as if in slow motion. Not only had I lost the straw but the 'eye' on the trailer drawbar had twisted.

6. Ah yes, Health and Safety issue. I fell backwards off a ladder as I was climbing up to get to the top of a stack of straw. A strange one that as normally I would go up crouching down on the old petrol driven elevator, we used for sending bales of straw up!!!

7. I was working in the field called 'Long Flatts', the one near the pig unit. I was driving the Ford 4000 with spring tine harrow on the back. I seem to remember that there were only one or two telegraph poles in that large field. Now I thought every square inch of soil should be worked. Spring tine harrowing is a job that needs to be done at speed. Yes, you guessed. I hit the pole nearest the pig unit. The pole survived. The three-point linkage on the Ford did not.

You may be thinking what a liability I must have been. In my defence I was young and inexperienced. Did I mention that before? It was a steep learning curve.

And so, my two-year tenure on this particular farm was nearing an end and a year at Agricultural College was on the horizon. I felt I had achieved a lot; I had earned my first proper wage which had enabled me to treat my mum and dad to a few things they never had.

I bought my first motorbike, a BSA Bantam 125cc. Which I actually seized up on the way home one weekend. Two wheels never did and never have suited me. I think I was cursed for having two wheels. I bought a windshield for it. The curved top type like you see on American police bikes. I had just finished fitting it under cover in one of the straw barns. I left it on its main stand and went in to put my gear on to go for a ride. When I came back out the bike had fallen over and cracked the screen right down the middle. I was miffed to say the least.

I had bought my first two cars. The first one I bought from my future Father-in-law for thirty pounds. It was a 1965 Ford Anglia which I later traded in for a 1968 Vauxhall Viva HB SL, A lovely car which I kept until 1977.

Ah yes, I met my future wife and readymade family.

Current music trend.......David Essex

Fashion......................Short skirts and Hot Pants

College

And so, to college. And another chapter opens up.

The local College of Agriculture was where I was heading for the final year of my apprenticeship. A year of intensive studying! The course would cover all things agricultural. With a view for many to go onto complete a second year in Farm Management.

The college was even closer to my home than the farm I had been working on for the previous two years. No more than ten miles away. However, the majority of the students, including myself, lived in.

Seems archaic now but the girls on the various courses run by the college were boarded in the upper floors of the main building. The boys were in two or three separate buildings, well away from them. Officially there was no visiting either way.

The room I was allocated was one of four on the first floor of a fairly newly purpose-built house. That's how I would describe it. It certainly looked more like a house than a college dormitory.

I was in a corner room. Pretty basic, but comfortable with a bed and a wash basin. There was no ensuite. Just a shared bathroom and toilet at the far end of this floor. The ground floor was exactly the same design.

Opposite my room was Harry, (H). In the front corner room was Frank. In the other two rooms there were a couple of lads over from Kenya. I'm not quite sure what course they were doing as they kept themselves to themselves very much.

On our floor there was also a small kitchen area with seating which was pretty good for whiling away the hours after a busy day in lectures. This area was shared by the whole house.

Myself, H and Frank formed quite a bond, and got into a few scrapes together, but mainly me and Frank.

More of that later. The business side first. The college was a working commercial farm, with pigs, dairy, sheep, machinery and a large amount of arable.

The structure of the days varied. Each of us had a rota whereby we would spend time working on each department on the college farm. This would normally be for a whole week of mornings followed by lectures later in the day.

I particularly remember the early starts, joining the head 'dairyman', Elwood, for milking. He was an amazing chap. I would imagine all good dairymen have this skill, but he amazed me by recognising which cow was which as they walked into the parlour purely from the markings on their legs. "That's Mirabelle, she wants X amount of food and will give X gallons today, That's Daisy, she wants X amount of food and will give X gallons" etc etc.

26

It was flipping cold standing on that parlour floor. Elwood would have major problems with his feet in the years to come. Coincidentally, many years later when he had moved on from the college to start his own livestock haulage company, I ended up living very near him and his family.

The upside of being on dairy duty was that you were first into the dining room for breakfast. I have never been a big fan of cooked breakfasts but the ones at college were amazing. This was the first time I had ever tasted properly fresh milk, straight from Mirabelle or Daisy that morning.

As it was a teaching farm some of the practices seemed a little labour intensive but that is how it had to be, I guess. The small pig unit was run by one man plus whichever students had been allocated on any particular week.

The dry sows were tethered just like on the farm I had worked on. Some of the farrowing accommodation was in a Solara building. Basically, a long 'lean too' design built of concrete blocks and asbestos roof. The pens housed one sow and her litter bedded with lots of straw. The pens were at least ten feet by ten feet from what I remember. In hindsight I suppose this was the forerunner of todays 'freedom farrowing'.

Fortunately for the college I was never released onto a tractor to perform any work out in the fields. Whether they had heard of my history on the arable side, who knows?

We did learn a lot in the machinery workshop, both practical and theory.

Sheep......I must have avoided. I hate them anyway. I remember having a go at shearing and dagging. The latter appeared to work on the principle that you let the sheep's back end get filthy and riddled with maggots before you catch them and cut it all off. The shepherd was an unusual chap. Passing on his knowledge was not on his agenda. He showed us once how he trained his new sheep dogs to lay down. We gathered around for his demonstration, expecting some secret dog whispering technique or something similar. How wrong were we? With a rope around the poor pup's neck and the other end in his hand he lowered the rope to the ground and stood on it. Then with a sharp tug on the rope and bellowing "lay down". The poor pup had no option but to lay down. Cruel man.

I am not sure I learnt a great deal to help me in later life, probably all down to me really. I remember I participated in the social side with more energy. Frank and I both joined the college rugby union team. I had always played at school and also for Hessle when I lived at home. I was the diminutive scrum half. Frank was a big prop forward. We played most weekends. On one occasion we had a game at our rival agricultural college, near York. We won. There were drinks and food afterwards which took us into the evening. Before we left, we had a look around the farmyard they had there, strictly out of bounds! We found the farm's combine harvester unlocked! One lad climbed up and grabbed the steering wheel and released the handbrake. The rest of us proceeded to push the machine from under its shed and down the yard and hid it behind another large building. I am sure they took their revenge when they visited us later in the season, but not as good a prank as ours.

Frank and I were as thick as thieves. We brewed our own beer, a really nice forty pint bitter kit. Some we sold to our fellow students, at a fair price obviously. One of our lecturers got wind of our operation but not before the beer had nearly all gone. We were both put on 'jankers'.

There was a solid wooden fence about one hundred metres long that needed creosoting. That would be our 'jankers'. We were given the brushes and drums of creosote. Now, not only was Frank a great laugh and a good mate he was also very enterprising.

"Listen Mike, there is no way we are painting this lot by hand. There is an old David Brown tractor parked up somewhere with a PTO driven paint sprayer on it. I saw it the other day."

Off he went, minutes later he rolled back up with the tractor. We attached what needed attaching. He drove and I sprayed. Twenty minutes later we were finished. We were seen by the farm manager, but he must have thought just how resourceful we were.

Frank was an accomplished tractor driver. As he was about to prove to me as I jumped up alongside him to return the tractor to where he found it.

"Do you know you can wheely one of these Mike?"

"No way," I exclaimed.

28

"It's easy," he said, "All the weight is on the back end, watch, and hold on."

With that he engaged third gear in the high ratio box and opened up the throttle full and let out the clutch. Instantly the front end lifted off the ground and we hurtled down the yard thirty or forty yards on the back wheels. Thick as thieves.

At times we had practical demonstrations on the farm so we could learn procedures we possibly had not come across before. One fond memory was being shown the intricacies of taking a sow's temperature. This involved the use of the old glass thermometers. We were given one between two of us. I was paired up with H. The lecturer explained in detail the method and demonstrated it with great skill.

"Whatever you do, do not leave go of the thermometer!" he said, very seriously.

H did like to chat incessantly and was easily distracted, by the way. He took the thermometer and carefully inserted it in the unsuspecting sow's rectum, at this point making sure he kept hold of

it. Now, he must not have had a really tight grip on the end. He turned to talk to me, about something totally unrelated, probably about his car or his girlfriend, at this exact point the sow gave a loud cough and a fart, and the thermometer disappeared, never to be seen again, not that day at least.

Needless to say, the lecturer in charge of us on that occasion was not amused. Mind you, I don't recall him ever looking amused when our particular group were with him.

On the subject of cars, a few of us at the college had managed to buy our own by then. I had my trusty Vauxhall Viva. Another lad had a really gorgeous looking Ford Escort Mexico Mk1. It was a deep metallic blue. An amazing car.

H had his Mini Clubman 1275cc. Anyone who has owned a Mini will know how much fun they are in any guise. I remember the gear lever was about two feet long and the dip switch for the headlights was on the floor next to the clutch pedal. On one occasion I was with him on the way back from the pub in Cherry Burton, a village near the college. There was a humpback bridge on the road back.

"Watch this" he said, as he pushed his foot down on the accelerator. We hit the hump and took off, all four wheels off the ground. He loved his Mini.

As I recall all the college 'activities', it seems little wonder that I ended up with just a pass in my final exams rather than anything higher. I have a list in front of me now and there are more 'leisure pursuits' documented than educational ones.

Many evenings were spent in the college bar drinking Coca Cola. The best evenings were after home rugby matches, singing risqué rugby songs.

Some evenings we would drive into Beverley, where we found Nellies bar. This is the oldest pub I have ever been in. It was owned and run by two ladies in their senior years. There was one small room with a pool table in the middle, or was it bar billiards? Whichever it was, by the time a dozen people were in there the room was full. The beer was kept in wooden barrels behind the counter, which was no more than a kitchen table. The amber nectar was tapped out of the barrels into a jug then poured into your pint glass. I remember a bowl of rather brown coloured water on the counter in which the ladies used to wash the empty glasses.

Well, I doubted Nellies would still be there, but I have just googled it. And low and behold it is still there, slightly revamped, but apparently has retained a lot of its quaintness. Well worth a visit. Now I am sounding like a Beverley tourist guide.

One evening a few of us went into Hull to a Nightclub called Romeo and Juliets. It was a proper sit-down nightclub. On stage that night was the late great Olivia Newton John. Why a group of agricultural students chose to go and see her I do not remember, however, she was excellent. I guess we probably fancied her.

And now I come to the most memorable and final recollection from my year at college.

At Christmas it was traditional for the students to put on a pantomime. All were involved in one way or another. It was to be performed on the final day of term before we broke up for Christmas, the day before Christmas Eve. The performance went well. It was Jack and The Beanstalk. Frank played the giant, he was the obvious choice. H played Jack. They were both sized to fit their roles perfectly. I think I was a stagehand or something like that. Afterwards someone suggested we should go through to the pub in Cherry Burton. So off we went. Now, at this point, I may remind you I am not a big drinker. Several pints of Watneys Red Barrel were consumed. Then when someone came up with the bright idea of buying a bottle of spirits each to take back to college to carry on the merriment, I just went with the flow and purchased a half bottle of Teachers Whisky. From this point my memory of what followed becomes a little hazy to say the least. Whether I drank the whole bottle myself I am not sure. What I do know is I became a victim of the evil drink. I was comatose. Very, very ill. It was also very, very messy. Fortunately, the lad whose room was directly below mine was kind enough to look after me, as I was to find out later. I think he may well have saved my life as I was so ill. At some point I managed to lose a false front tooth that was on a palate. It never turned up and I never found it. I have a suspicion the lad who looked after me may have kept it as a souvenir, he was a little strange. What else he got up to while I was unconscious, I dread to think.

The following day, Christmas Eve, it was the end of term, and everyone went home. I have no idea how I got home. I cannot remember anything from the moment the Teachers Whiskey kicked in. I missed the Christmas of 1974...... I've not touched Teachers Whiskey since.

We returned to college in the New Year. Six more months of study interspersed with all the other leisure pursuits. It seemed to pass by so quickly. My final exam results were pretty average, just a pass. But it turned out that my best subject was Accounts. And wait for it, my worst subject was Pig Husbandry!

Despite this slight drawback I secured a job managing a small pig unit with just eighty breeding sows.

Current music trend…...Queen

Fashion…..................Jumbo collars

Interlude

My writing has just been interrupted by Christmas 2022. This is the first Christmas families have been able to get together since the Covid pandemic. Although it should be a time to celebrate, the world seems to be in turmoil. Russia has invaded Ukraine. Prices are escalating because of this. The NHS is at breaking point. Things can only get better, surely.

Once you have retired, the approach to any breaks change. I think anyone who has retired may understand what I mean. I had similar feelings earlier on in the year when we went abroad for a two-week holiday that had been deferred several times due to the pandemic of 2019. We all remember that. Travel was restricted, and of course the dreaded 'lock downs'.

So, to my point. When I was working, and a holiday of any description was approaching I would be rushing around like the proverbial 'blue arsed fly'. Just to ensure everything on the farm was up to date, vaccinations, bedding and a long list of day-by-day instruction for the chap who would fill in for me while I was away.

By the time it came around to the holiday I really felt as though I had earned it. Even if I was to be away for the odd two-day bank holiday, I would prepare things in the same way.

I did not expect the feelings to be so very different just because I had retired. However, they are. Of course, the holiday itself was great, as holidays always are. And Christmas this year was as good as any. It was the build up to them both that I missed, I guess.

I am sure I am not alone feeling this way once in retirement mode. I suppose it is the whole lifestyle change, post-retirement. For instance, weekends are no longer weekends if you see what I mean.

Waffling interlude over. Back to the story.

First real job

I was employed as the manager of an 80 sow breeding unit. Small by any standards, but as good a place as any to start my career in the pig industry. The unit was part of a relatively small arable farm dedicated to grass and cereals. The farm was situated on the outskirts of the market town of Knaresborough. This was my first real job earning real money. Well, £40 a week.

For the first couple of months, I lodged with a couple in the town. They owned a cafe in the town centre where I would go for a meal after work. There was to be a 'tied' house that went with the job, but this became available later.

The pig unit was only small. The farrowing and dry sow accommodation was all under one roof. The weaner accommodation was in a building adjacent to it. These weaner pens were a similar design to the unit in Kilham. The only difference was that they were sat on slurry channels, so in theory there was no scraping out to do. Unless, of course, the pigs decided to muck in the sleeping areas.

All the sows were tethered in both the dry sow house and farrowing house. I know I should be able to remember how this unit worked regarding numbers farrowing per week. However, when I do the sums, it looks ridiculous. It was definitely only an eighty-sow unit, which by my reckoning if I was weaning at five weeks old and achieving 2.2 litters a year for instance. I must have been farrowing no more than four sows a week. Which means a maximum of twenty farrowing crates in the building.

Feel free to check my calculations. It is sounding more like a hobby farm now.

There is no wonder I had time on my hands to help out on the arable side of things. I got involved with harvest, silage time, and I painted the boss's house outside. I even re-painted an old Massey Ferguson tractor. Yes, it seems like I was haunted by the little red beast from Kilham.

There was not a lot of corn grown on the farm. I remember loading small bales of hay onto trailers by hand. There was a fair amount of silage made. This was made into a silage clamp.

Basically, an enclosed area in half a building. The outside wall of the building at one side and a wooden sleeper arrangement at the other side would contain the silage. Trailers would be tipped outside the barn, and it was taken in and heaped up with a tractor with a 'buck rake' (a wide fork), that was fixed on the three-point linkage. As the clamp was built up higher and higher it was necessary to erect a steel ramp to drive the tractor up. The action of driving over the cut grass obviously compressed it to form it into silage over a period of time. I am no expert, but that was the theory.

That was a hot summer. It must have been, as on numerous occasions the boss's young daughter would come out to help wearing just shorts and a bikini top. Oh yes, and it was the summer I accidentally caught the boss's wife sunbathing topless. It was an accident, honest, I was looking for my boss.

I was only at this farm for one year. The house that was provided with the job was a little grim. A problem with damp that seemed impossible to cure. As by this time my future wife and her two children had moved in with me, I decided it was not suitable and it was time to move on.

My, or should I now say, our next move was to prove to be a long term position and a memorable one.

Current music trend......Abba

Fashion..................... Well, I still had my flares.

Job for life?

The farm was situated in a small hamlet near Boroughbridge. There were two other smaller farms in the village and just eight houses, including the farmhouses. We were given a three-bedroom semi-detached house which was owned by the farm and was just across the road from it.

When I started working there as manager there were just 120 breeding sows. It hadn't been running for long. It was actually a new venture for Jim, my new boss. Himself and his younger brother had taken on the running of the farm from their father. It had mainly been sheep and arable prior to Jim deciding pigs were the future.

At the time I started working on the farm there were two men working the arable side along with Don, Jim's brother.

I could see from the start that Jim had ambition and drive and a definite plan for where his business was heading. For this reason, I knew this was going to be our life for a good few years. And as things turned out, I was right. Twenty-Five years in total.

So, brace yourselves readers as I attempt to recall all the highs and lows of those years that shaped my future.

The breeding sows were housed in a pretty standard design 'Brian Thomas' building. There were four rows of stalls where the sows were tethered. Behind each row there was a muck passage which was scraped out a couple of times a week with a Honda scraper, similar to the one at Kilham. The sows were fed by hand in a long, glazed trough which ran the length of each row. A large header water tank sat above the end of each row, which enabled the flush water drinking system as was still popular then. Twice a day they would be given water. A lot less was known then about the importance of individual animals' water intake needs. So, this system was acceptable. At the top end of the building there were four boar pens. The freshly weaned sows were held in the two middle rows at the end near the boars to make it easy to run them to the boars for service. Once the sows were served, they were moved into the stalls vacated by the heavily pregnant sows that had been moved into the farrowing accommodation.

The unit was run on a three-week weaning system which once again meant we were weaning, serving and farrowing every week. With a small sow herd this was the best and easiest method.

Although the breeding sow housing was modern the farrowing house was a conversion of an old building. It was a bit of a mish mash design but made best use of the space. There was one long row of farrowing crates. Then up a slope there were another eight crates, four either side of the top of the slope. All the pens holding the crates were raised up, so the muck fell through the steel slat directly behind the sow and was then scraped out by hand and wheel barrowed away. The plus side with these farrowing pens was that they had a large front creep area with a lid with a hole in it which housed an infrared heat lamp. The piglets were bedded with wood shavings.

Once again, all feeding was done by hand.

The weaner accommodation was one long building. Another 'Brian Thomas' design. There was a feed passage which ran the full length of the building. As you walked down the passage on your left on the outside wall were manually operated louvre windows, on the right were the pens which held around twenty piglets per pen. These pens were totally slatted with the 'diamond type' expanded steel and sat over individual slurry pits. Each pen had a sluice gate on the outside of the building enabling the slurry to be emptied into a larger channel running the length of the building. This channel was open and about six feet deep and eight feet wide. To access the sluice gates, I had to walk along a plank which also ran the full length of the building.

There was a full height asbestos partition between each pen and then a door in the passageway between each three pens in order to control the temperature of the different age groups. Heat was supplied by calor gas heaters hung on chains above each pen. These piglets were all hand fed into hoppers at the front of each pen. They were fed a specialist high energy piglet creep feed for the first couple of weeks then moved on to a less potent feed later.

At one stage when we had too many pigs to accommodate in the weaner house Jim purchased three purpose-built weaner huts.

These were made of fibreglass, with a domed roof that slid back over the hut to allow the filling up with pigs and emptying. These three pens were actually placed directly over the slurry channel. I seem to remember they were not the best accommodation I have encountered.

From the weaner accommodation they would move into the grower house. A straw based building with a covered sleeping area and outside muck passage but within the building. It was a high roofed barn so straw bales were stacked on the top of the pens. We could store a lot of straw in there and also in the next stage building which was a similar design that housed fattening pigs. The actual process of getting the small bales in there was pretty labour intensive as you could only access the buildings from either end, and not actually drive into the buildings. The buildings were probably about eighty yards long. So, a trailer of straw would be parked at the end of the building and one by one the bales would be thrown onto the platform which was the roof of the sleeping area of the pens. We would form a chain, so to speak, throwing the bales along to whoever was stacking them. It was a pretty dusty, hot job but a chance to bond as a team.

Jim was in the process of increasing the size of the herd from the day I started working for him. He was already putting up another large building which was to house farrowing sows at one end and breeding sows at the other.

The sows were still to be tethered in stalls and the farrowing sows in crates with tethers. The farrowing pens had front creep areas for the piglets to sleep in. Heat for the piglets was provided by infrared heat lamps as in the older building. There were two rooms with twelve crates in each.

At the other end of the building there were four rows of sow stalls with thirty in each row. The whole building stood over slurry channels, which meant no mucking out by hand. At the end of each channel was a sluice gate. The slurry flowed from here to the holding pit at the end of the weaner house. These sluice gates often stuck. It was in an attempt to lift one of these sluice gates that caused my first back injury. Jim offered to give me a hand to lift it out. He had a great idea. He hooked a short chain onto the loop at the top of the sluice. With a steel bar pushed through the

38

other end of the chain we stood, bent over, at either end of the bar.

"Right Mike, on three we lift."

"One …...two…....three."

He lifted on three, I was just too slow, and I was jerked down. I must have torn something in my spine. If I am honest, it has never been the same since. Lesson learnt. Don't let your boss interfere.

And so, as the herd grew more staff were needed. 'Geordie' was employed to help me run the breeding side of the unit and 'Billy' was taken on to look after the growers and fatteners. We were a good team. Although on one occasion myself and Geordie did have a confrontation. Over something ridiculously trivial as I remember. It all blew over quickly.

At this time Jim was ploughing money into the expansion of the unit. However, in some areas investment was a little less evident. For instance, for mucking out the grower and fattening house we had a small dumper, the type you see on a building site, with a rubber blade on the front. The blade was on a steel frame that was raised and lowered by hand with a lever. The dumper needed to be started with a cranking handle. Some of you will know how painful starting one of these can be if your timing isn't right. It was a matter of holding up the decompression lever with one hand and turning the crank handle with the other. Once you had some speed up you needed to move the decompression lever over and quickly pull the crank handle off the starter shaft. If you didn't time the process to perfection the handle would kick back with the risk of damaging your hand and wrist.

The dumper was also used for pulling the homemade pig trailer around on the unit. There is a tale involving that later.

One of the routine jobs in the breeding herd was the castrating of the young male piglets. Someone in their wisdom had decided that meat from the male animals was tainted if they were left intact. This forced producers into the situation whereby all the males needed to be castrated.

This was a two-man job, although we did try a framework once that you were supposed to be able to suspend the piglet in, allowing it to be a one-man job. This wasn't really a brilliant idea

39

as two people were needed anyway to catch the piglets in the farrowing pens.

So, one of us would hold the piglet between our legs! The piglet would be upside down and held by its back legs. The one wielding the scalpel, invariably me, would pinch just behind a testicle and squeeze gently so the flesh became tight. A delicate incision was made about half an inch long just cutting the skin so the testicle could be pushed out. It was then just a matter of pulling it out, cutting the vas deferens, but pulling the blood vessel until it broke. By pulling the blood vessel, rather than cutting it, the 'shock' to the vessel reduced bleeding to a minimum. Once both testicles were removed a squirt of diluted Dettol was applied. The poor little things would scoot around the pen after, scraping their wounds along the concrete floor. The piglets were about ten days old to make it easier to find the testicles. Interestingly the farm cat always knew when it was castrating day. She loved them.

This procedure sounds barbaric now and thank God it was decided to be totally unnecessary in later years.

Another rather gruesome idea our vet had at the time I shall now describe.

If you are squeamish, skip this part!!

In an attempt to prevent a persistent neonatal scour, which I assume at the time there were no vaccines for as there are these days, this is what was suggested.

The first idea was to mix the piglets scour with water and feed it back to the pregnant sows, the idea being that the sows would build up an immunity and pass it on to the piglets. This seemed not to work. The next suggestion was as follows. We should feed the dead piglets back to the pregnant sows! Yes, you read that correctly. Now you are thinking, "How on earth would that be achieved?" Well, brace yourselves. Jim was advised to purchase a mincer. It was an industrial size. The size you would find in an old butcher's shop. We found it was not possible to feed a whole piglet into the opening in the top of the mincer. They needed to be reduced to smaller portions. So, an axe was used to split the carcasses in half and then into smaller pieces. I feel nauseous describing this. It sounds absolutely ridiculous now. I refused to do it after a short trial. I think Jim realised why I was against the practice when I demonstrated it to him.

Once again, thank God things moved on and vaccines were developed. Vaccines which actually worked and caused less stress to both pigs and pigmen.

Squeamish readers can carry on reading now.

The whole farm was so well set up. Not just the pig unit. As I say, we milled our own feed which was pelleted in the very expensive pelleting plant Jim had installed. The man who looked after the mill was also in charge of the workshop. 'Chuck' was his name. He could make anything in his workshop. From the sow stalls to the body of the double deck forty-foot stock trailer. Later on, as the farm expanded, he also made outdoor weaner and farrowing huts with the help of his young assistant 'Leo'.

Not only did we mill our own food, we also had our own blower wagon which was operated in the early days by Jim's brother Don. At some point later Don decided to break away and set up a farm on his own with his new wife. The first farm he bought was out near Knaresborough. I remember going stacking straw there and I am sure we must have kept some of our pigs there as Don had complaints from the villagers about the smell. Even though there had always been a farm there for many years before 'town' people decided they wanted a taste of country living.

When they moved from there a few years later they bought a derelict farmhouse. The farm had several acres of arable land and a small pig unit. Coincidentally I ended up managing that one later on when Don decided to rent it out to another pig producer.

I recall going to see the farmhouse before Don and his wife started the renovation work. It had not been lived in for many years. There were holes in the roof and Pigeons had taken the house over for roosting. There was a sad looking Harmonium in one of the rooms covered in Pigeon excrement. Funny how certain images stay with you and others do not. It must have cost a fortune to refurbish the place. However, it looked amazing when it was finished.

Things seemed to be moving along at a pace on Jim's farm, both at work and my home life.

Shortly after moving we decided to get married. February 1978. So, there I was, twenty three, married with two children. We had a really nice house with a big garden. We even kept some geese in the paddock and at one point had a Toggenburg nanny goat and a little white goat called Gladys. We had a large grey rabbit too. I had built a large hutch for the rabbit. Quite often when we left the hutch door open Gladys would go in it and go to sleep. Unfortunately, the rabbit got run over, not that it was the goat's fault. My wife was taking the children somewhere and had them loaded up in the car, in the driveway at the back of the house. As she pulled forward, she felt that she had run over something, so she backed up to see what it was. It was the rabbit, poor thing. It had chosen the wrong time to have a wander around.

We also had a good sized vegetable patch where we grew all sorts. Blackcurrants, Raspberries, Strawberries and various vegetables. I used the fruits to make wine. It was very good. The things you do.

At this point in time the farm must have been making a lot of money. As well as the investment in the farm John also put in a private tarmac drive up to our house as originally, we shared a drive with Chuck who lived next door.

Within a couple of years Jim decided to expand the farm yet again. This time with a completely new four hundred sow unit. All the buildings would be from the firm 'Brian Thomas'. The unit was really well laid out. The dry sow house was in two halves. At one end down one side and half the length of the other side were boar pens. The rest of the space was taken up by two and a half rows of sow stalls. The sows were tethered again. It would be a few years before this was banned. This end of the building was for the newly weaned sows for serving. They were kept at this end until they were four weeks pregnant so that they were near the boars and could be checked easily for returns.

The other half of the building was where the pregnant sows moved to for the rest of their gestation period. There were four rows of stalls with two feed passages. The boars were on straw. The sows at both ends were on partial concrete slats, so they laid

42

on solid concrete and the back ends on the slats. All were hand fed and watered with the usual flush water system from a header tank above each row.

Next to this building was a row of farrowing houses. There were two rooms of ten pens. Then a gap, then three rooms of ten pens. The farrowing crates were imported from France. 'Galvepor' was the firm if I remember correctly. They were a bolt together design. To save space in the design of the building there was just one central passage. The sows faced the passage. Moving sows in was not the easiest of jobs but once they learnt it worked pretty well. The creep areas for the piglets were at the head end of the sows but at one side. It meant that one creep lid, which was made of thick asbestos, covered two pens. Thinking back, it was a slight compromise. The lid held one calor gas heater. This was the first time we had used a gas heater in the farrowing houses. They were to be used in the next stage also, where the weaned piglets were moved to. The gas heaters had a foam filter on them which as you can imagine collected lots of dust and had to be cleaned regularly. These gas heaters had no 'fail safe' of any description on them.

Next to the farrowing houses there was a large gap to allow tractor access. Then at the far end there were four flat deck buildings, to hold the weaned piglets, with two rooms in each building. Each room had four raised pens, two on each side. The pens sat above a shallow slurry pit. The pens were totally slatted with expanded steel, the 'diamond' type. Each pen held about thirty pigs. A plastic feed hopper formed the front of the pen.

In line with these rooms was the first of two grower buildings. Both buildings were one third slats and two thirds solid concrete as a laying area. There was one central feed passage in each. In the first building the pigs were floor fed. In the second building they had ad lib hoppers which were filled by hand. The pigs were moved from the first building to the second at about 30 kilos.

The last building on the unit was one long fattening building, separated into two rooms. Once again, these pens were partially slatted and partially solid concrete. The feeding system was an 'EB' feed system again, the same as the one on my first unit. The ventilation system was supposed to be the latest innovation in pig building ventilation, 'High speed'. It turned out to be hopeless.

43

There were fifty two pens in there in total and for a long while you could go in there and have to squeegee fifty two pens of slurry before you could feed. We actually ended up feeding by hand so as to encourage the pigs to eat in the sleeping area rather than use it as a toilet. We tried all sorts to stop them. However, nothing seemed to work. Years later when I moved on to yet another expansion of the business the building was converted into farrowing accommodation.

That was the only weak point of the unit. If you looked at an aerial view of the unit, it was all set out perfectly in line. A track ran the full length in front of the unit to enable the feed blower access. One long passage ran along the front of the buildings, for feed barrows and pig movements.

All of the sow stalls on this new unit were made in our own workshop by Chuck and Leo. Chuck was a perfectionist. I remember helping him setting out the stalls before they were concreted into position. He would stand at one end of a row of stalls, crouched down looking along the line. I walked along the row, and he would get me to tap any stalls that were slightly out of line with a mallet. When the builders came in later to put the final layer of concrete down, he would be there keeping an eye on them to make sure none of the stalls were moved.

I was to have sole management of this unit such was the trust Jim now had in me. It was to be a three man unit. Nick joined my team along with Dave who was totally new to farming. Dave and his family had just moved back from New Zealand. He was in graphic design originally. We were a good team and got on really well together.

When we first started stocking the unit with the new young breeding stock, before Nick and Dave had started on the unit. The building work was still in progress with only the sow house totally finished. It was a very wet time of year.
Therefore, the whole area was a messy, wet building site literally. The new breeding stock had been housed on the old unit temporarily. Now, as I mentioned earlier, the only equipment we had for transporting animals any distance was the dumper and the homemade stock trailer. So, Geordie and I set on to move the first twelve Gilts over to the new unit. We decided as it was so wet, we would be better to reverse across the field as we would not be

44

able to turn around to back up to the sow house, plus it would mean less manoeuvres. I think Geordie was driving, however that is immaterial. As we got nearer the new unit our progress got slower. The wheels on the dumper started spinning and we were soon bogged down. Only halfway across the field, about eighty yards short of the new sow house. Time for 'plan B'. We had no option but to drop the tailgate and walk the gilts from there. Let me say that pigs behave a lot differently than sheep or cattle when shown a vast open space to explore. Sheep would have just stood there, then once you get one pointed in the correct direction the rest start to follow. Not so with pigs. It was a big field, a very big wet field. We dropped the tailgate and stood either side ready to direct them towards their new accommodation. The first two gingerly stepped slowly down the tailgate, sniffing as they went. Then with a bark and a grunt they were off, sprinting away. Before we had a chance to lift the tailgate back up the remaining ten gilts decided to make their own bid for freedom, off they went in twelve different directions. It was their first taste of freedom. You would be amazed how much ground a pig can cover in a short space of time. Before we knew it, they were dots on the horizon, (a slight exaggeration), but the field was about twenty acres. So, what should have been a quick little job became a rodeo as we went about the task of gathering them up and walking them back one by one to their new home.

The next challenge was getting them into the new stalls and tethering them up. Fortunately, there were gates at both ends of the rows of stalls which meant once we had finally got all twelve gilts into the building, we could enclose them within the area with the two rows of stalls. These young gilts had never seen a stall before or had a tether around their necks. The method we employed was as follows. A small amount of food was placed in the feed trough in each stall. Pigs will do anything for food. It was then a matter of encouraging the gilts to walk in one at a time. At this stage the gilts were very nervous as you can imagine. Once one had walked into a stall, I would get alongside it while Geordie stood behind it, stuck his knees in its backside and braced himself by holding onto the side of the stall. I would then put the tether around the gilts neck. I then had to climb out with Geordie still braced in position, I would bolt onto the stall a 'U' shaped framework which stopped the gilt pulling back and pulling its tether off. Geordie could then escape. We only had the twelve frames so moving gilts over was a slow process as it would take a few days for the gilts to settle down enough to have the frame removed.

All the breeding stock for both units was bought from PIC. One of the biggest breeding stock suppliers at that time. I am pretty sure we may have bought these new gilts in various stages of pregnancy to speed up the process of getting into full production. Before long, it seemed, we were fully stocked and started farrowing. The farrowing houses worked well once we sorted out the easiest way to move the future mothers into the farrowing crates. The thing with pigs is that you can train them to do almost anything for food.

At some point we decided we needed extra farrowing space and also extra weaner piglet space. It was decided the easiest way to create this space was with self-contained portacabins. One had ten farrowing crates in it set out in a 'herringbone' style.

The extra weaner piglet space was in the form of three portacabins. Each had two rooms. Each room had two, three tier cages giving us twelve cages in each room. Each cage only held ten pigs. There was a fibreglass trough in the front of each pen. The floor of each pen was metal slats with a fibreglass tray underneath each one to catch the muck. This tray sloped into a fibreglass drainage channel which served the cages at either side. The rooms were heated with the gas heaters. This idea of the cage system was another experiment as we reduced the weaning age to fourteen days. This is something else that is now illegal. It was an attempt to get extra litters per sow per year. However, it was decided that pushing the sows this hard was detrimental to their health and longevity.

By this time, we had stopped having to castrate the male piglets so that was one less task. The processes we did use within the first twenty four hours of the piglet's life, and still do to this day, was to cut out the piglet's teeth, take off part of its tail and give them an iron injection.

At birth piglets have eight very sharp incisor teeth, four at the top, four at the bottom. If these are not removed, they can damage the mother's teats, and each other. The teeth were cut out at gum level with what was basically a pair of wire cutters. The same cutters were used to cut part of the tail off. I would leave about an inch of tail on. This was done to prevent 'tail biting'. Pigs tend to be a little cannibalistic if there is any trace of an injury on one of their own. Therefore, if the tails were left at full length there would be more chance of the pigs starting to bite each other when bored. As the industry learnt more about pig behaviour it was decided that introducing 'manipulable materials' into the pig's pens would reduce boredom. Boredom and overcrowding being the main causes of tail biting. Also, to help prevent the tail being an open wound a cauterising tool was introduced rather than using the wire cutters. Many producers have now also introduced teeth grinding into their piglet treatment routine which has done away with the wire cutters completely. Just the tips of the incisors are ground off. These introductions were many years away from where I am up to in the story.

Many of the systems and building designs were 'cutting edge' at the time, although now when I look back, they seem archaic. There are several instances to illuminate this. In the farrowing houses the use of the gas heaters proved to be disastrous on one occasion. Remember the sows were tethered. The partitions between each pen were only about sixteen inches high and slotted into steel channelling on the outside wall and the front of the pen. The channelling was held in place with bolts through the strips of punched metal slats.

One morning I arrived at work and walked into the first room of the two farrowing houses. All was well but as I could smell something acrid. I quickly ran into the next room. It was carnage. The room was full of smoke and several partition boards were smouldering away. A sow had slipped its tether and backed out of its crate. It had wrenched up several partition boards but also upended several of the gas heaters and lifted up some of the slats. I turned the ventilation fan up to full speed to get rid of the smoke. The sight was horrendous. Two of the sows were burnt as the gas heaters had landed close to them. They were asphyxiated as were another four of the ten sows in the room. Almost every piglet in the room had suffered the same fate. The lucky ones that escaped the acrid smoke had slipped through the missing slats and fallen into the slurry pit.

Unfortunately, the slurry pit was half full and piglets can only swim for a short time. With the smoke clear from the room and the smouldering wood extinguished my attention turned to rescuing the piglets from the pit. By this time help had arrived and we set on with the rescue. It was a matter of removing the manhole cover outside the farrowing house so we could pull the sluice gate up on the slurry pit. Hoping that as the slurry flowed out the piglets in the pit would flow along with it so we could grab them and lift them out. The pigs were in a sorry state. Some were dead. We pulled about forty pigs out. Only half would survive. Another lesson learnt.

The flat deck building was another area that at the time seemed ok. The ventilation system in here drew air in from outside through gaps in the outside walls then up through the slats and through a mat of domestic type fibreglass insulation which formed the ceiling of the room. The fibreglass was held in place on a wooden frame by chicken wire netting. There was no way of cleaning the fibreglass which got very dusty. The problem was exaggerated when it inevitably got wet when we pressure washed the rooms. The slats over the slurry pits in these rooms and in the grower buildings were mild steel and not galvanised which reduced their lifespan. Within three years of the unit being built they were beginning to rot. Therefore, a close eye had to be kept on these. Pigs have a habit of finding any weak spots in any situation, and once they find something they tend not to leave it alone. So, if they found a loose corner of a rotting slat, they could soon peel it back off the wooden frame. Not only are they very inquisitive, but they are also very strong. We did have to retrieve the odd pig or two out of the grower house slurry pits over the years.

Although the unit sounds like it had so many flaws it was what was available in pig building design then. As the years went by, designers and pig producers alike discovered better and safer systems. Modern units today are virtually 'bomb proof'.

Supermarkets demand producers keep up to very high standards now both from a welfare point of view and the safe use of medicines.

But for now, we are in the late 1970's, and learning as we go.

There was another large farm a couple of miles up the road from ours. Jim was good friends with the owner. The workers and their families and the farm owners' families used to get together for events during the year. The Christmas meals and parties / discos were brilliant. We had inter farm cricket matches and the odd barn dance. Things like this made us feel appreciated. Jim even gave us all gifts at Christmas, and when he was still growing potatoes, we were allowed to take whatever we needed within reason. All this exceptional generosity came to a halt later when 'belts had to be tightened', and more money was ploughed into yet another expansion of the business which meant extra staff was needed. Towards the end of my time managing this particular unit for Jim, we had four pigmen on each unit, two men running the workshop, two men on the arable side, a full time wagon driver, a full time secretary, and a 'personal assistant' for Jim, who needed to spend more time in the office from now on. All this growth had happened within the first six years of me starting working for Jim. There was more to come. However, before I move on to the next stage, I would like to relate to you a few memories of those first years.

Let's start with the vehicles Jim owned. When I started, his family car was an Austin Allegro, the one with the square steering wheel. His other car which we used as the farm workhorse was a Morris Marina estate with a tow bar on it. We used this for taking cull sows to the market in Thirsk, among other things. One time I set off with five sows, in the old wooden horse box, heading to Thirsk. The horse box was not in the best of conditions. When I got to the market and unloaded the sows there were only four. They had been so unsettled enroute that one had jumped out over the tailgate. We found it eventually in a field just off the A168. She seemed quite happy with her new found freedom, although it was short lived.

49

As the business started making some money Jim bought himself a short wheelbase Mitsubishi Shogun. I remember mobile phones were being developed then. Jim had one in this vehicle. The whole thing was the size of a concrete block. He had to take it in his house each night to charge it up. His family car was now a Volvo estate. He had also acquired a rather nice old Porsche 911.

At this time, I still had my 1968 Vauxhall Viva, but I steadily worked my way through several old cars in those first eight years at Milby.

In 1979 we had a son. A little baby for his two sisters to spoil. It must have been at this point we decided we needed a car that suited our family size. I bought a Mini Countryman. The one with the wood trim around the windows and room in the back for a push chair. I remember replacing the whole front end because the wings were rusty. I bought a full fibreglass bonnet that hinged to open. I then hand painted the whole car.

Spooked

I used to do late night checks on the pig unit, just to make sure everything was ok especially if there were sows farrowing that may need help. Two vivid memories, in particular have often cropped up in conversation subsequently, are as follows.

It must have been late October or maybe into November. We had stopped up late watching 'The Fog', a horror film. It was later than normal for my 'night check'. So, I put my hat and coat on, opened the back door to find there was thick fog outside. I couldn't believe it. There was no street lighting and no security lights around the farm. I was a little scared to say the least. A walk in the fog I do not mind, but a walk in the fog after watching that film. Spooky.

The other 'night check' related story is far less plausible, and to this day I don't think any of my family believe me when they hear this. On this occasion we had not been watching a film, I had not been drinking and the weather was fine, it was in fact a star lit night. I need to say that as otherwise you readers would also doubt what you are about to read.

The distance from our house to the pig unit was about four hundred yards. Once across the road which passed through

Milby, there was a track that actually was a 'right of way' which went through the farm and on to the next village of Kirby Hill. I had completed my 'night check' and began to head home. There was no one else around. As I came alongside the feed mill, I realised there was something above me. Totally silent, directly above me there was a massive disc shaped object. It was lit up like a Christmas tree. If I was to liken it to anything now, I would say it was a cross between the Millenium Falcon from Star Wars and the main spaceship in the film Close Encounters of The Third Kind. Now I know what you are thinking, and I don't blame you for doubting me. It seemed very low, only a matter of maybe forty feet above me. It seemed to hover over me for a while, then moved away slowly and silently as it rose into the night sky and disappeared.

Why did I not report it?

I do not know

Did anyone else see or report it?

Not to my knowledge.

However, I know what I saw. And maybe it was seen by other people. And maybe it had the same effect on them, as in not reporting it! If only mobile phones with cameras had been invented then. But that begs the question, would a camera have worked? It takes some explaining I know, maybe whatever it was had been attracted to the heat source coming from the pig units. It is a memory that has stuck with me all these years.

I've just stepped back from part two of the book as my daughter has reminded me of a story I should include involving our son. You know when you have a boy and you want him to be able to do everything 'NOW'? Like, riding a bike or kicking a ball or swimming even. Or in this instance, flying a kite. I had bought him/us a stunt kite. At the back of our house there was a long grass field that would be ideal for flying the kite. The field must have been about one hundred metres long. The grass was not too long at the time so should not impede the kite flying. I gave Christopher the kite to carry and we walked up to the far end of the field. When we got to the end I explained what the plan was.

"Right, you hold onto the kite while I walk back to the other end of the field with the two control handles unravelling the strings. Once I get down there, I will shout, and you throw the kite up into the air as hard as you can."

"Ok daddy."

The kite stood a foot taller than Christpoher. He was just three years old, standing there with stunt kite in both hands in his brown velour shorts and matching top. (Was I expecting too much?) I set off back down the field walking backwards as I let out the two lengths of nylon string. It's a long way one hundred metres. When I got to the end of the field, I pulled on the strings to take up the slack. I caught Christopher unawares and he dropped the kite.

"Pick the kite up" I shouted, he couldn't hear me. "PICK THE KITE UP", he still couldn't hear so I set off back up the field to where he was.

"Hold on tight to the kite and when I say throw it up you just throw it up as hard as you can."

"I was going to, but you pulled it out of my hands daddy."

"Hold tight until I shout."

"I'm trying daddy."

So, off I went to my end of the field again. I picked up the control handles. I waited for the breeze to pick up. Christopher was hidden by the kite in the distance. The breeze picked up.

"Throw it up now," I shouted....................." Throw it up now," I shouted again. He heard that time and threw it up, but not hard enough. "Pick it up and try again,"............"Pick it up and try again."

Poor Christopher picked the kite up and tried one more time. Without success again. He had had enough and left the kite laid on the grass and started walking back down the field towards me.

"What are you doing? We can make it fly."

"I don't want to play."

We gave up. It would be at least another seventeen years before our next venture into stunt kite flying.

The builder we used on the farm for a few years was a 'Jack of all trades'. It was quite evident from the sign writing on the side of his van. It read something like, "Joe Bloggs, builder, plumber, electrician, decorator ". All these skills seemed absent when you saw the quality of his work. I once saw him laying concrete blocks at various angles to finish a wall off rather than cutting them. On another occasion he was fitting some kitchen units in our house. He had turned the electric supply off for safety reasons but then stood scratching his head wondering why his electric drill would not work. My wife had to point out what the problem was.

It was him I sold my Vauxhall Viva to. A few weeks after he bought it, he split a tire and expected me to refund him for it. The tires had been on the car for five years! Cheeky bugger.

The cars

On the subject of cars. In the eight years we lived at Milby working on this unit we went through several old cars. First the Viva. Then the VW Beetle, followed by the Mini Countryman. It turned out that the Mini floor was porous! When the roads were very wet, water would come in through the floor. When we went uphill the children had to lift their feet up. Going downhill the water flowed to the driver's footwell. Our next car was a blue Vauxhall Viva HC. My first Viva was the HB version. This was a decent car. Unfortunately, when the MOT came around it needed some welding. It was in a garage in Boroughbridge. We got a phone call the morning we were expecting to collect the Viva. There had been a fire at the garage. There was good news and bad news. Only one car was burned out as it was high up on the car lift. The bad news was that it was our Viva. Apparently, the welding that was normally damped down after being done had smouldered and set on fire overnight.

The Insurance was paid out without a problem fortunately. It had been a really reliable car up to that point.

We bought a Renault 6 next. That was a seriously bad buy. Two days after we bought it the alternator failed. Of course, it was 'bought as seen', so I had no come back on that. Later that year when it was due its MOT, I had the rear jacked up so I could check the wheel bearings by grabbing the wheel and giving it a waggle. To my horror it was possible to pull the wheel out away from the car, still attached to part of the subframe. It was an instant 'write off'. The worry was we had recently been over to Cheshire on the M62

After the Renault we bought a Simca 1100. It had a horrible bright orange velour interior. We only had that for one year. I think we decided we needed more space, so I bought a Hillman Avenger estate. This car suffered its terminal fate on the way back from Cheshire one weekend. We were just outside Harrogate when our son decided he was " bursting for a wee ". I pulled in at speed into a rough layby and smashed down into a large pothole. The result was a broken chassis. So, it was goodbye Avenger.

For our next car for some reason, we went small again. We decided we could afford a brand new Mini Saloon. It turned out we could not realistically afford the payments at that time and had to sell it back to the dealer. This was not the wisest move I have made in my car ownership life. I think I panicked and thought it was the best thing to do.

At this point we went back to second hand cars. It is also at this point that my memory fails me. I know at some point we had a Ford Fiesta in a horrible metallic rose gold type colour. The only thing is I am not sure whether this is where it fits in the list. I am pretty sure it was the car we had when we made the move to the next farm that Jim bought and wanted me to manage. It is immaterial really, but I was trying to keep things as accurate as possible. I know we must not have kept it for long as I do remember the next car that we had after we made the move. It was another Fiesta in dark blue.

So, that's our car history for now. There are quite a few more to follow later though. An interesting point is all of these models are classic cars now.

On the arable side of the farm the pig men used to get involved at busy times. Bringing straw in at harvest time and also when it was potato harvest time. Jim grew quite a large acreage of potatoes for quite a few years. This stopped abruptly when prices plummeted. I think there were so many farmers growing them. I think the final straw was when our wagon driver arrived back at the farm with his full load. The factory who normally took them had refused them for some spurious reason. He had tried taking them elsewhere, but nobody wanted them. Reasons ranged from the potatoes being too big, too small, too dirty. So that was it, Jim stopped growing them.

I used to enjoy harvest. The small bales were dropped from the back of the baler in eights. I would follow on with a flat eight loader. It was an attachment on the front loader of the tractor. It was a matter of driving into the bales and operate the hooks on the frame of the machine with a hydraulic lever. These flat eights were heaped five high and collected later with another tractor and machine that picked up all forty bales and either loaded them onto a trailer or took them straight back to the farm, depending how close the field was to the farm. All of the straw was stacked inside the pig buildings.

The pig men never did any land work such as ploughing etc. This was all done by the two arable men, Chuck when not busy in the workshop and Ronnie.

The dogs

Just down the lane from where we lived in Milby there was a canal and river which ran parallel with each other. We used to walk our two dogs down there most days. The path in between the two led all the way down to Boroughbridge. Our two dogs at that time were my wife's Labrador cross Lurcher, Jason, which she had had when I met her, and a Labrador cross puppy, Blue. Both were lovely, all black, dogs. One day I was down by the canal with both of them, off their leads as always. To this day we do not know for sure what happened. One minute they were sniffing around in the undergrowth and the next minute Blue, the young pup, shot out and started running back home, Jason came out of the grass looking dazed. He was staggering and could hardly walk. I picked him up and carried him home. When I got to our house Blue was laying in the hallway gasping her final breaths on the front doormat. We rang the vet, and he came out straight away. When he arrived, Blue was dead. Jason was very ill but still alive. The vet took him away to his surgery. A few days later we got the call that Jason had died too. The vet found traces of Potassium Phosphate in his blood. He reckoned that fishermen sometimes use it. The reason Blue had died so quickly was because she had run home, and it had gone through her system so quickly.

We heard that someone else had a dog that had suffered the same fate around that time. We informed the police, but nothing was ever discovered that proved what had caused the deaths.

The cat

On a lighter note. We had acquired a lovely looking tabby cat which we called Benny. He used to love going out hunting in the paddock next to our house which led onto open fields. On the other side of the paddock though was the main road which ran through the village. He would spend all day outside hunting. When he was ready to come in, he would scratch at the utility room door until one of us heard him. He would trot off upstairs and lay in his favourite position on the landing where some central heating pipes ran under the floorboards.

One evening he never turned up at his usual time. At first, we thought he must still be busy hunting. As it got later, we began to worry so I went out calling him. There was no sign of him. The following morning, he had still not returned. We searched around but we could not find him. We thought the worse, that maybe he had been run over on the road and maybe thrown in a hedge back somewhere. We resigned ourselves to the fact that he had gone for good.

The children soon got over the loss of Benny, like they do. A couple of years passed. And here comes something else that you may find hard to believe. One evening my wife and I were in the kitchen making tea when she said,

"There is something scratching at the door out there."

"I didn't hear anything."

I went and opened the utility room door, and unbelievably there was Benny as bold as brass looking as fit as ever. He came strolling in as if he had not been away. He gave us both a sideways glance, meowed, and trotted off upstairs to take up his favourite spot on the landing. We never did find out where he had been for two years. He was not one for giving away his secrets.

The boat

One day I was down by the canal when I spotted a small motorboat drifting along. It was on the side of the canal that was not easily accessible, but I managed to scramble through some brambles and over a barbed wire fence to get near the boat. Fortunately, it was just drifting slowly with it being on the canal. As it snagged on a branch that was sticking out from the bank I reached out and pulled it in. There was a rope in the boat, so I tied it to a tree. It was only a small boat, maybe four people would fit in it. There was a fishing rod and a waterproof jacket in it. It had an inboard motor. A Morris Minor car engine. It seemed a strange situation. Which became even more strange. I went home thinking, "I'll keep that". My honest wife had different ideas. Insisting that I ring the police. So, I did. They seemed disinterested and simply said if nobody claims the boat in six weeks, I would have the chance to take on ownership but it was a matter for Waterways to sort out.

Apparently, nobody had reported it missing and once the Waterways Agency were involved, they told me if I wanted it, I had to buy it off them as it was on their property. Needless to say, I did not buy it. The one thing I  never understood was how you could lose a boat like that and not know. What crossed my mind was, had someone fallen in the canal and drowned with there being some belongings in the boat. I never heard anything else about it. I could have lazed many hours away in that.

Current music trend......Bucks Fizz

Fashion.......................... Permed hair (Kevin Keegan)

57

More expansion

In 1994 Jim bought another farm. It was on the outskirts of the small village of Crayke, on the foot of the Hambleton Hills. The farm was surrounded by about sixty acres of arable land. There was a large five-bedroom farmhouse with a large garden. The house and the unit were at the end of a track with nothing and no one around it. There was a working pig unit on the site which had fallen into disrepair which was to be rebuilt completely. The farmhouse, unfortunately, was in a similar state. I think the owner and his family were so busy trying to keep the farm going that things had just got on top of them.

Jim asked me if I wanted to move up there and manage the new venture. We would be starting from scratch, basically a 'greenfield' site once all the old buildings that were not needed were demolished. I jumped at the opportunity to move. I was always ready for a challenge and the setting was idyllic.

There was a lot of work to do before any building work could even begin. As Milby was only a thirty minute drive from Crayke my family and I continued to live there until we started buying in the breeding stock.

The first job was to sell off all the sows that were on the unit. All the sows that were within two weeks of giving birth we kept on the farm and sold them later. The same applied to the sows that were suckling litters. As you can imagine this was a slow process as Jim had bought all the stock on the farm. This included the growing pigs too.

All the pregnant sows were sold to one big unit near Wakefield. I remember sorting out their recording cards meticulously so that the new owner would know exactly when each female was due to farrow. When we loaded the sows up onto our wagon and I made sure I put them on in the order they were to farrow.

I went along with our driver with the first load. I remember the unit was at the back of a large housing estate. It was apparently built over some old coal mining works. The unit was massive. Most of it was under one roof. We pulled in and backed up to the loading bay. Eventually a couple of men appeared. I explained how the sows had been loaded up in some sort of order to make it easy for them to house in that order. I handed the cards over explaining that they were for this particular batch. One of the men took them off me and said,

"We won't need them" and threw them onto a nearby muck heap! So, I guess they just used potluck to decide as and when they would move them into the farrowing accommodation, if they even bothered that is.

While we were there the men showed us the main building. It was enormous. There were farrowing crates at one end and sow stalls at the other. There was a raised walkway which ran the full length of the building, I guess to observe the pigs and maybe the multitude of young people that were there seemingly endlessly sweeping the passages. A surreal experience, I have never seen anything like it since and not sure I want to.

Once we had sold all the old breeding stock the new farm building work commenced. A whole new building was to go up to start with. This was to house pregnant sows and gilts at one end with three boar pens. There would be two hundred and twenty stalls at this end. In the other half of the building there would be five farrowing rooms and eight flat deck rooms just across a passage from them.

We decided we would move away from tethers at this stage. This meant the sow stalls had to be longer so as to allow the sows to lay comfortably. Although it must be said they were still pretty much enclosed. A simple plywood board slotted into some channelling at the back of the stall to hold the sows in.

The farrowing crates had a simple 'gate' on the back that had three positions so that it could be adjusted to fit different length sows. All of the stalls and crates were made in our workshop at the farm in Milby by Chuck and Leo.

Thinking back now this was all a massive undertaking, an enormous investment at the time.

The aim of the unit was to be what is known as a 'multiplier unit' for PIC, the company we had used before. We would supply females to various units around the country. The males would be sold for bacon.

Jim had found a different builder to undertake this project. This new man had his own labourer and were a team that had worked together for a few years. They were employed mainly to build the foundations of the buildings that were to go up. The concrete block walls and slurry pits etc. On a good day Bill, the main man, could beat anyone at laying blocks. On a bad day, he may not turn up, if the demon drink had got the better of him. But when he and his mate were in the mood there was no stopping them and he was a perfectionist, surprisingly. One week in particular they didn't turn up for two days. It was the beginning of the week. On the Wednesday when they did come to work Geoff, the labourer called me over. He put his hand in his trouser pocket and said,

"Mike, look at this, I had a bit of luck on the horses this weekend." He pulled out a large wad of ten-pound notes.

"Three thousand," he said. No wonder him and Bill had been on a drinking 'binge'.

So, work proceeded at a pace I was up there just about every day. As was our digger driver Martin, who coincidentally lives in the same village as me now.

He was and is a character. A self-confessed 'expert' JCB digger driver. I worked most of the time with him. Digging out the channels for the slurry pits and digging trenches for various drainage pipes. A couple of memorable jobs we were both involved in deserve a mention.

The first involved digging a trench to put in a main nine inch slurry pipe which was about forty metres. Jim had explained how much fall was needed on the full length of the pipe. To ensure we got this correct it involved the use of a spirit level as we laid each section of pipe. Maurice was adamant he knew exactly what he was doing. The number of pipes he had put in over the years was not worth thinking about. Jim left us to it. I just followed Martin's instructions. By the end of the day, we had the pipe in place but not back filled the trench. The following morning Jim came to have a look at how things were going. It had rained heavily overnight. He stood over our handy work looking bemused. One end of the pipe was totally submerged, the other end was above the water level. He called us over, not too happy. There was far too much fall on the pipe.

"How on earth have you managed this? You've got about two feet of fall, it only needs three inches." I had no answer, however I remember the response from the 'expert'.

"Nowt to do with me, Mike had the spirit level." That is Mike the pigman not Mike the experienced pipe layer. Jim laughed it off and suggested we take the full length of pipe out and lay it all again, which we did.

The other memorable job was, once again digging a trench but this was a lot deeper under what was to be the service house.

Because the new site was on several different levels on a sloping field it was necessary to bring in many loads of hardcore to level it up. However, half of the unit was on much higher ground than the area that was to be the farrowing and pregnant sow building. This meant channels to carry drainage pipes were so deep that Maurices digger was at full stretch to reach the bottom. It must have been twelve feet deep at least. I was in the bottom of the channel basically tidying it up as Maurice dug it. I was in, what seems now, a ridiculously dangerous place. Although the soil was mainly clay and in theory solid, the sides were vertical and there was an obvious risk of the whole lot collapsing in on me. Maurice suggested we make some pit props as a precaution. So, he found some old timber and constructed a few crude frameworks which we jammed in the position where I was working. To be honest I doubt they would have helped if the walls had collapsed. It was another one of those jobs that I look back on and think I should have refused to do. Similar to the propcorn one earlier on at Kilham. However, more seriously risky jobs were to be undertaken by me in the coming years, and up to my retirement I was still taking risks.

Over the ensuing months the new unit was beginning to take shape. The site was levelled up, slurry pits were dug and blocked and rendered and we were ready for the shell of the building to be erected. This was a specialist job to be supplied and erected by Brian Thomas again. The whole of the buildings come as a giant flat pack. Because of this, erection was pretty swift as the men on the job were so experienced. Within a couple of weeks the whole shell was up and roofed. The next job was for Bill and Geoff to get in there to lay the concrete for the sow stalls and passageways and put the glazed feed troughs in place. Once that was done Chuck began bringing in the two hundred and twenty sow stalls. Leo was back in the workshop at Milby fabricating the last few stalls and then starting on making the farrowing crates. I helped Chuck get the stalls into position. Ever the perfectionist each row had to be perfectly lined up before being secured. Lengths of steel tubing held the stalls together, but it was Bill's job to put the final layer of concrete down to secure the feet of the stalls in place, without moving them.

The timing of buying in the new breeding stock needed to coincide with the new building being ready. To facilitate this Jim rented a large barn in the next village. We used it to house our new gilts and a couple of boars so that we could start serving. By doing this the first gilts to farrow were ready as soon as the new farrowing houses were ready.

By this time my family and I had moved into the farmhouse, it was April 1984. All three children were settled into their new schools. My wife came up to help me serving the gilts. She was determined to pet the boars. I had to point out to her that it was not too good an idea as when they get to be two hundred and fifty kilos you don't want them nuzzling up against you.

The set up in the barn wasn't ideal but it was all we had, and it meant we would have pregnant gilts ready to move in once the new buildings were ready.

The whole process worked out perfectly and we began to stock the unit while the next stage of buildings were being completed. They were, the service house, a grower house and two fattening houses. The whole site was to be enclosed inside a perimeter fence to deter uninvited visitors as it was to be a 'High Health Unit'. Noone was allowed onto the unit without showering first and changing into the unit's clothes. We had a purpose-built shower block, bait room and office. All the bulk feed bins were inside the unit, but wagons blew the food into them from pipes outside the fence. The site was chosen as a High

Health Multiplier unit because of its position more than four miles away from any other pigs.

Eventually there would be four of us working on the unit. Nick, who was to be my fist assistant manager came up from the unit we had worked together on at Milby. He was to live in a caravan on site. It was handy for him as he was closer to home rather than living in Milby. We were a pretty good team, which helped the unit run smoothly and efficiently.

Once all the buildings were finished we were able to stop using the barn in the next village and move everything to the new unit. We had built a gilt service shed purely for gilts. All the serving was natural, not artificial insemination as it is on most farms these days. This meant in total we had about fourteen boars on the unit. Nick decided we would name them all after classical composers. Beethoven, Handel, Bizet etc. Most of them were pretty timid, apart from Rimsky Korsakov. On one occasion he almost got me.

The way the service house for the sows was laid out was as follows. Looking at it from the end. On the left side was the sow pen sleeping kennel, then there was a muck passage, then a step up into the individual feeding crates. In front of them was the feed passage and on the right hand side were the boar pens. The sows were in groups of eight in each pen. There were eight pens. To feed them they would run into the crates and the hinged gate at the back would hold them in. The sows that were due to be served were left shut in when they had finished eating. To allow them closer boar contact prior to being ready for serving we would run a boar up the feed passage. This contact helped the sows to ovulate.

Rimsky Korsakov was at the far end of the building, and it was his turn to have a run. I opened up his pen gate and walked up the passage to where the sows were. I was only half way up the passage and he came racing up alongside me. Fortunately, I had a plywood moving board in my hand. As he passed, he swiped his head to one side and gauged the board with his tusk and tossed me up onto one of the feeding crates. A fully grown

63

boar can have tusks at least three inches long and very sharp. It was a lucky escape.

The pregnant sow accommodation, farrowing houses and flat decks were built over  slurry pits. The sow service house, grower house and fattening houses were all straw based. The service house was as I described earlier. The grower house had one central passage with pens on either side. The sleeping area was covered and had a hinged lid the full width of the pen and about three feet deep. Outside of the sleeping area was a muck passage about eight feet wide. This building was on an automatic feeding system with the centreless auger just the same as on the first unit I worked on in Kilham. The very same systems are used to this day. It is simple to use and very reliable hence its longevity.

The two fattening houses were a pretty standard design too as in, feed passage, sleeping kennel, muck passage. Then there was a central wall and on the other side a mirror image.

One fattening house was dedicated to housing the gilts and the other was for boars. We separated the gilts from the boars at weaning as we were going to be selecting the gilts to sell as breeding stock. The boars were sold for bacon. All four of the buildings at this end of the unit were mucked out with a small Weideman articulated tractor. Because of the various levels on the site at the far end of these buildings was a wide concrete pad then a vertical drop of ten feet. This made mucking out a little precarious. If you overshot, you were in trouble.

These buildings were also used for storing straw on top of the kennels just like we did on the other farm.

Once we were in a position to start selecting the gilts a selection officer from PIC came. This was a two man job. I assisted with this, with a view to taking over the selecting eventually. The selection process was very rigorous. Any gilts that did not meet the criteria were sold as bacon.

We were looking for:

Conformation
Fourteen accessible teats
Good strong legs
No dipped shoulder

Conformation included basic body shape and condition. They should have good hams. Not too thin.

The teats had to be well formed, we were looking out for inverted teats which would be no use. When we started, we accepted twelve accessible teats. However, one of the PIC sales representatives told a customer that PIC gilts could suckle fourteen piglets. Word soon got around and from then on fourteen teats became our aim.

The legs were a particularly important part of the selection process. No gilts with swollen hocks were selected. They never recover as they grew older and looked unsightly.

Dip shoulders is a hereditary defect. It appears as it sounds. If it is present, it is quite pronounced. One of our biggest customers was really keen on this. One load we sent her had one dipped shoulder in amongst them. Somehow it had slipped through. She rejected the whole load of fifteen gilts saying most of them were dipped. PIC let this go as she was our biggest customer.

When the gilts were selected, they were all ear tagged with numbers and all the details recorded. This made allocating the gilts to our different customers a lot easier. The 'cream' of the selected gilts were allocated to our best customers. Some were less picky than others. All the gilts going off the farm were top quality, with just minor differences.

Demand for our gilts was high and as we reached full production, we started exporting selected gilts to Germany. These were lucrative orders as they were to stock new units.

I got the chance to travel over to Germany with a load of gilts. That was a great experience. We loaded up and drove to Hull. We sailed over to Rotterdam on North Sea Ferries then drove through to Germany. When we arrived at the main farm we were welcomed as if we were long lost family. One of the PIC selection officers met us there. We thought we would have to unload the pigs for a quick turn around. But we were invited into the manager's house where his wife offered 'kaffee und kuchen'.

65

I was to travel back home with Alan, the selection officer. We stopped off at a large beer and wine warehouse on the way back to the ferry terminal where I bought my first ever crate of Grolsch. The bottles with the flip off wired tops. As we left the port we had a near miss with another ferry. It was nothing major, just a close shave.

Nick, my assistant manager, had taken this trip earlier in the year. We later found out that the highlight of his trip was the stop the driver made in Amsterdam. The story involved a rather pretty lady, scantily dressed, a lift, and Nick's bad timing. He blamed the lady's perfume and her close proximity. I don't think it sounded as though he got value for money.

The business was flourishing. Prices were high and our gilts were in great demand. We were growing our own corn for the pig feed and milling our own feed. The only feed we bought was the specialist creep feed for the baby piglets.

Yet another venture Jim entered into was supplying heart valves intended for human use. These were taken from the progeny of a small herd of Yucatan pigs he had bought. They did not produce many piglets but apparently the heart valves were much sought after.

Current music trend ......Simply Red

Fashion............ Moving towards shell suits

The farmhouse we were living in at this point was perfect. It had four double bedrooms and one single bedroom. Downstairs there was a large lounge with a bow window looking out onto the front garden and an open fireplace. The kitchen was enormous, we had a big table with benches in there that my dad had made for us. It comfortably seated eight and left plenty of room to walk around. From the kitchen there was a large sitting room that we used as our 'everyday' room. This had an open fireplace too. Both the kitchen and the sitting room had original oak beams. A door from that room led into what we called the den. We had a mid-sized snooker table in there but at Christmas we used the den as a dining room. The large lounge was a room we only used at Christmas. There was room for a nice big tree. With the fire lit it took on a classic seasonal look.

The original layout of the house was not ideal. The back door opened into the large lounge straight from the messy backyard. There was only one staircase which was opposite the front door. This meant the lounge was a main thoroughfare. Jim very kindly agreed to alter the layout to suit us. So, we agreed to block up the existing back door and build a utility room on the back which would house the boiler, chest freezer and washing machine. From here a new door led into the kitchen. From the kitchen we put in another staircase. This made a lot more sense. The backyard was tidied up and gravelled and double gates put on to keep children and dogs in. There was a big garage that would easily hold three cars. From the garage we could access the front garden and a stable, which did come into use for a while. There is a story there involving a pony. Later.

Once the whole house had been decorated, and a new kitchen put in and central heating we decided to splash out and carpet the whole house apart from the bedrooms in a quality carpet that would last a long time. So, we covered the den, sitting room, lounge, hall, both staircases and landing with the same carpet. It cost a lot but was well worth it.

The front garden was big enough to play cricket on and there was another more formal piece of garden at the side of the house. The whole house was surrounded by fields that belonged to the farm. The field behind the newly built unit sloped fairly steeply down to a stream. Great for the children to play. Bordering the field to the East of the farmhouse was a hedge that had so many different types of trees and bushes. It was hundreds of years old apparently. Fortunately, it had never been decimated by machine hedge cutters. Through this hedge was a footpath that led up to the Hambleton hills.

Along the banks of the stream there was a plentiful supply of dead Elm trees. I used to go down there with the Weideman and pull them over and drag them back up to the house. I had bought a chainsaw to cut them up. We never bought any coal for the fires.

This sloping field was fantastic for the children for sledging when it snowed. To the left of the track which led around to the back of the unit was a grass paddock where we put the pony for grazing.

Lady

Ah yes, the pony, Lady was her name. For some reason we adopted it as we had a stable. Our son took to riding her as if he was a natural. The neighbouring farm down the track used to host a gymkhana in summer. He entered and won a couple of rosettes, dressed in his trainers, jodhpurs and woolly jumper. Not the classic kind of dress for a country gymkhana.

Now, do not entertain any thoughts that we were a 'horsey' family. We took the pony on as a favour. I will go as far to say I don't even like anything equine. But for my son's benefit I entered into this new experience. I would groom her with him. Muck the stable out and walk Lady out to graze in the paddock. One day I even took her out for a ride across the fields and back through the village. To this day I still do not understand what pleasure people derive from this pastime. Lady seemed to take a shine to me and would follow me in and out of the stable to the paddock, and back again at night…...normally! You know what is coming?

It was early evening one summer's day. We had been to the railway station in York to pick up our youngest daughter who had returned from working in Majorca as a travel rep. We decided to stop off at the pub in Crayke to celebrate her coming home, just a couple of beers. As we drove up the track to the house my wife said,

"You best get Lady in."

"Let me out here then."

"You will need her halter!"

"No, it's ok, she just follows me."

So, I got out of the car, and they drove up to the house.

If you remember my encounter with alcohol at college, you know I am not a hardened drinker. I had only had a couple of pints but was a little tiddly. I swayed and stumbled to the gate of the paddock.

"Come on Lady."

And up to the gate she came trotting. I opened the gate, and she walked out and followed me. I was pretty chuffed that I had been able to train her to do this with no halter.

At the end of the short track that led down to the paddock you join the track that leads up to the house. Turn left to the house and her stable, turn right to go to the road to Crayke. I turned left.

"Come on then, let's get you to bed."

I turned to make sure Lady was still following, only to see her cantering off down the track. She had decided to turn right. I chased after her, shouting expletives, I am sure. It was no good. I couldn't catch up. The last I saw of her she had turned right into a field and disappeared into the dusk. I ran back up to the house, out of breath.

"Did you get Lady in ok?"

"No, she has run off."

You can imagine how the rest of the conversation went. We both went down the track in the hope that she may have stopped to graze. She was nowhere in sight. My wife rang the police to let them know about our escaped pony. I think the Constable on the other end of the phone giggled. They said they would let us know if they heard anything.

The following morning the police rang and informed us that Lady had turned up at a stable a couple of miles away. Apparently, she had picked up the scent of some stallions on the farm and made a bee line for them. She was in season unbeknown to us.

The very kind stable owners brought lady back in their horse box later that day. They saw the funny side of it too.

Pets

We had a bit of a menagerie while we were living in the farmhouse. As well as the pony we have two dogs, a cat, a budgie, a hamster, four geese and Danny the duck. Danny was hilarious, he used to come in the house to tell us he was hungry. He also used to terrorise our son when we played cricket or football in the garden. He would chase after him pecking at his heels.

The budgie was in a cage hanging from one of the beams in the sitting room. At that point in our lives both my wife and I were smokers, not heavy smokers, but smokers all the same. It turned out that the budgie was addicted to cigarette smoke. We realised this years later when we moved from the farm into a house we bought, and we gave up smoking. The budgie died. Nicotine deprivation?

The geese used to live in the paddock with Lady. They had a shelter I had made but not one that I could close them into at night. One morning I went to see how many eggs there were and found three of the geese dead and one missing. There was a trail of feathers across the paddock. It was obvious a fox had had them. Apparently, that is what they do if there is a group of birds. Take one and kill the rest, so I was told.

Our two daughters soon got to the age when they could hit the road. For the eldest I bought an old Mini Clubman. A lovely little car but the paintwork was a little rough. I decided to repaint it. I didn't have access to a sprayer, so I hand painted it with a brush. It didn't look too bad when I finished. Our youngest daughter decided she wanted to be on two wheels, so I found her Honda MTX 50cc. It also was a little rough around the edges, so I stripped it down and repainted it with white Hammerite paint. She rode it a few times but decided that arriving at the sports club discos looking dishevelled after removing her crash helmet was not quite the look she wanted. So, I sold that bike and bought her a Fiat Panda which I also brush painted in white. It sounds like I was a bit of a dab hand, but I wasn't. I think we eventually had problems with the Fiat, and I ended up changing it for another Mini. It was a lovely Mini 1000 in a pale metallic blue.

Our son decided he fancied a motorbike once he saw his sister had one. He was only ten. I decided to let him have a go and bought a cheap moped that needed some work doing on it to get it working. There was a decent sized grass area he could ride around just outside our front garden. He was messing about on it one Saturday afternoon. I had left him to it. He seemed to be doing ok. Suddenly he appeared in the kitchen holding his arm

70

with tears streaming down his cheeks. His arm was swollen between his elbow and his wrist. It was definitely broken.

"We will have to take him to A&E," my wife said, "Quick, get the car started!"

Apparently, my response was that she would have to take him as we were deep into the World Cup, and there was a match I did not want to miss. My wife has never let me forget that I stayed, and they went.

Our financial situation was far more stable at this point having had a substantial pay rise and with my wife working at the local building society. So, we decided to buy a new car again. We decided on a new Vauxhall Astra. It was when they went for the curvy design following on from the old angular shape. We traded in the blue Fiesta for this one at the Vauxhall dealership that was in Easingwold. That is now yet another housing estate.

There are so many things that happened all those years ago it almost sounds untrue as I put them on paper. Two such tales involve each of the two Minis. One slightly more distressing than the other.

One afternoon my eldest daughter needed a lift to work in Easingwold, so I decided to take her in her black Mini. In we got and set off bouncing down the track, like you do in an old Mini. At the bottom of the track, before you turn left to reach the T junction on the main road, there was the gamekeeper's house. He worked for the lord of the manor who owned a lot of land in Crayke. Anyway, he and his wife were standing outside and as we passed by, they were waving really enthusiastically. We waved back. We turned left and headed to the T junction. I braked at the junction and a black tail appeared on the windscreen, wagging. Teddy, the young black cat had ridden all the way down the track on the roof. Silly us, no wonder they were waving so enthusiastically. He let us

71

pick him off the roof, put him in the car and take him back up to the house.

The other incident involved the blue Mini and a series of coincidences. I needed to go to York for some reason. I took the blue mini. I parked in the car park we always used. Went off and did my shopping. When I came back to the Mini it would not start. The battery was flat as I had left the radio on. I jumped out to pop the bonnet. As I lifted the bonnet something caught my eye as if something shot off across the car park. I never gave it a second thought. I was more concerned about getting the mini started. Fortunately, someone in the car park had some jump leads and helped me start it. With that I set off home.

When I arrived home my wife asked

"Have you seen Priscilla?"

Priscilla was yet another cat we had at that time. She was not a well cat.

"No, I've not seen her, however!" ….......

I went out to the Mini and popped the bonnet. We couldn't believe what we saw. All over the rocker cover on top of the engine there were obvious scratch marks. So, what must have caught my eye when I popped the bonnet in the car park was the cat. It had obviously travelled all the way to York, thirteen miles, holding on to the rocker cover.

So, like a devoted, pet lover I returned back to the car park in York to have a fruitless search around to see if she was still there. Needless to say, she was not.

Now back to the unit. Despite having to cope with the differing levels of the site we had managed to lay it out well, in a way that made pig movements nice and easy. The weaned sows would stay up in the service house until we had checked them for their three week returns. They would then move down into the sow house and were kept in the order they were due to farrow. They all had individual sow record cards with all the relevant dates and information on. The feeding in here was pretty basic but we did feed extra in the last three weeks of pregnancy to aid piglet growth. We also pregnancy tested the sows at around the fourth week of pregnancy. Although the machine we had was not as hi tech as the ultrasound scanners we use today. This one was about twice the size of a Weetabix. It had a flat probe at the end, two lights, one green one red and an on off button. A small amount of 'light oil' was applied to the probe then it was a matter of putting the probe on the sow's skin just in front of her back leg at the side just above the back of her udder. Pointing towards her womb. If you got a red light, the chances are she was not pregnant. If you got a green light, she more than likely was pregnant. If you got a green light and a buzzer you had hit the jackpot and she was definitely pregnant, maybe!

I used to go in to pregnancy test the sows at around two in the afternoon. They all seemed to be settled at this time and were invariably all laid down which made the job easy.

From the sow house they were walked into the farrowing house via the central passage that separated those rooms from the flat deck rooms. We used to give the sows a scrub down with a Savlon solution to clean them and tone the skin up.

At weaning, when the piglets are taken off their mothers, it was a simple matter of running the piglets across the passage to the appropriate room. It was at this stage that the piglets were sexed, boars together and gilts together. The sows were backed out of their farrowing crates and walked up the slope to the service house. We weaned on a Thursday. Within four to five days they would cycle and be served again. Each sow received two serves twenty four hours apart in an attempt to catch them at the peak of their cycle which normally only lasted thirty six hours. And so, the whole process started again. If the sow stayed pregnant, she would farrow in one hundred and fifteen days. Or as an easy reference, three months, three weeks, three days.

By this time Jim had employed another chap to work on the arable side in Milby. His main tasks involved muck spreading and slurry spreading. He used to come up to the new unit at Crayke to empty our slurry pits when necessary. He was eager to please and had a good heart but was a little slow as in 'one sandwich short of a picnic'. One day he came up with the slurry tanker. It was a big twin axle tanker that would hold about eighteen thousand litres. He was spreading the slurry down in Crayke somewhere, on a neighbouring farm. This meant he was up and down the track negotiating the narrow exit past the gamekeeper's house at the bottom of the track. I was in my office when I noticed him on his way back up the track. To my amazement there were only three wheels on the tanker. As he turned in to go round the back of the unit to the slurry store, I opened my office window and waved, shouting to attract his attention. He stopped and jumped out of his tractor.

"Hiya Mike, what's up?"

"Err, Corky, you appear to have a wheel missing."

"Oh, bloody hell." he said. "When has that come off?" He was now quite animated and frothing at the mouth, like he did. While he stood scratching his head staring at the hub where the wheel should be, the gamekeeper pulled up outside my office in his pickup truck.

"Does this belong to you?" He asked, pointing to the enormous wheel he had in the back of his truck. The wheel had come off as Corky drove past the gamekeeper's house and rolled into his backyard. He must have travelled all the way up the road into Crayke, spread his load of slurry and come all the way back without noticing. If anyone had been in the way when the wheel came off, they would have been badly injured. The moral of the story, keep your nuts tight.

On a more serious note, while I was working at the farm at Crayke I had the unfortunate task of dismissing one of my team. He was only a young lad and had a bit of a temper on him and got frustrated easily if the pigs he was moving did not move quick enough for him. He had been reported to me a couple of times for mistreating them, for which I had warned him to curb his behaviour towards them.

He failed to control his temper and I spotted him being violent with a group of pigs he was moving. I was not prepared to let him carry on working for us and I dismissed him straight away. He wasn't a bad lad but had issues that he needed to sort out. But not whilst working on my farm.

Our youngest daughter was now working in an independent travel agency in the nearby market town of Easingwold. She encouraged us to have our first ever foreign holiday in the October of 1988. We decided to go to Malta for two weeks. Just the three of us, my wife, our son and myself. We flew with Dan Air. Do you remember them? Smoking was still allowed on planes back then. When we landed in Malta, we had our first experience of when you feel that Mediterranean heat as the aeroplane doors open, and you stand at the top of the steps. It is still a great feeling to this day every time we go abroad.

We collected our luggage from the carousel and headed out of the arrivals area to meet our rep. She welcomed us and pointed to the transfer coach. I grabbed our cases expecting them to be going on the coach with us.

"It's ok, just leave your cases, they go separately in that van."

She pointed over towards an old looking Luton box van.

I reluctantly let go of our two cases. We boarded our coach, at the same time keeping an eye on our cases. We then watched as they were tossed haphazardly into the back of the transit. The man loading them up then pulled down the roller shutter door on the back of the van and went round to the driver's side, jumped in and sped off. As he bounced over a couple of bumps the roller shutter door  jerked up and opened with all the cases shifting around as he disappeared into the distance in a cloud of dust just as our coach set off.

Malta looked half-finished as we drove across the island to our hotel at the Northern tip near the ferry terminal that serves Malta's sister islands Gozo and Comino. Paradise Bay was the name of the hotel. To our great relief the transit van with our cases was already parked outside the hotel when we arrived. Cases were intact. Time to chill with a bottle of Cisk lager.

The hotel was lovely, the staff were genuinely friendly. Malta is a beautiful island. All the buildings were built using the locally quarried limestone blocks. We drove past one of the quarries whilst over there. They actually cut the building blocks directly out of the quarry walls using massive disk cutters.

It was our son's birthday while we were there and also the birthday of the daughter of a couple we met. After the evening meal on the day of their birthdays one of the waiters came over with an enormous birthday cake and everyone in the restaurant joined in singing 'Happy Birthday'. Needless to say the children were embarrassed.

The cake was delicious, the cream was like nothing I have ever tasted before or since.

This was the first of many foreign holidays. We were hooked. We returned to Malta a few times over the next few years.

One of the dogs we had, while living at Crayke, was a little Border Lakeland cross called Sammy. We got him off the gamekeeper at the bottom of the track. He had spent his first few months in a kennel outside at the gamekeepers. He was very timid. However, we soon earned his trust, and he settled in as a family pet although still a little nervous, but always happy to see people. Sometimes slightly too happy. Our eldest daughter had a fiancé who visited regularly. We would see him coming up the lane in his car and tell Sammy,

"Mark is here".

Sammy would rush to the back door to be let out to greet Mark. On one occasion as Mark opened his car door Sammy jumped in, sat on Mark's knee and did a wee. He was always pleased to see Mark. Despite this over friendly greeting Mark still married our daughter in the summer of eighty six.

As well as Sammy we also had a Chow cross called Henry, and Bruin our lovely Alsatian cross Collie. Bruin moved with us from Milby as it happens. If you google the breed, Hovawart, it's the spitting image of him. He was as soft as a brush. However, people who didn't know him used to cross the road when they saw him coming.

Henry was also a softy, he had the look of a slimmed down Rottweiler.

Poor old bruin began having trouble with his sight as he got older. We had to take him to a specialist Vet in Holmfirth to have cataracts operated on. It improved his sight for a short while but soon deteriorated again. From about nine years old his back legs began to weaken. It was his hips really, a trait associated with Alsatians. He was also having occasional accidents with his lack of bladder control. We decided we did not want to put him through any more operations at his age and so tried to make his last months with us as much fun as possible.

We were in the middle of quite a cold spell, with snow and strong winds. Bruin was not bothered by any weather. We thought he was nearing the end of his life so I decided if he was going to go, he may as well go doing what he enjoyed most. There was a virtual blizzard blowing outside.

 My son decided he wanted to come with me. So, we dressed for the weather and set off with Bruin for our long, possibly final, hike. We went all the way across the fields following the footpath up the hill towards Yearsley, then we came back along the main road towards Crayke and cut through some woods battling our way through brambles and bracken. When we got home, I dried Bruin off, and he took up his place in front of the fire. We were all prepared for the worst the following morning. However, to our surprise he greeted us with a new spring in his step. It was as if the long hike had rejuvenated him. He lived another four months before he finally passed away.

We had nine happy years in the farmhouse on the new unit. Our son had started playing football for a nearby village team. I was playing for a team in the next village on a Saturday afternoon and Sunday mornings. Sunday morning matches were a little hit and miss, in so far as we never knew how many players would turn up due mainly to the after effects of alcohol the night before.

One Sunday morning we turned up at our usual meeting place where we would set off from for an away game. We had a match against a team in York. They were renowned for being a really good team. We set off with only seven players thinking maybe some more would turn up at the York team's ground. They did not. We had to make a decision. Play with seven, any less and the game was officially cancelled anyway, or go home, or see if we could borrow a player or two from the opposition. They told us no, in no uncertain terms. Our captain decided we should play, and we all agreed. Seven versus Eleven, what was the worst that could happen? We played very defensive and left one lad up front ready for any long balls being cleared from the onslaught. As it happens it was one of our most enjoyable Sunday outings. We managed to stifle the York team's attacks pretty well. They only managed to score seven goals past us, and we even managed a consolation goal.

When our son was eleven, I started coaching the team he was in and carried on with that bunch of lads for six years through to the under seventeens league. It is fair to say we were not the best team in our area, but we were not the worst. We had some really good players. There was a mutual respect between me and them. The majority of the lads stayed for the whole six years. They were like an extension of my family.

I've just been reminded about the two fires we had whilst we were living up there. Well, one was more of a smoulder.

The first one was on what is now a big grass roundabout. I think this one may have been down to Martin. Remember Martin? 'Nowt to do with me' Martin. He had been making a heap of hedges that had been pulled up and some old wet big round bales of straw. When he lit it, the bales didn't burn straight away and in fact it looked as though they had gone out with just the hedges having burnt.

It was about two in the morning when the flashing blue lights woke us up. My wife opened the bedroom window to ask the policeman who was standing just outside our front garden what the problem was.

"You have a fire love."

She looked out of the window to the left and to the back of the house.

"Oh my god Mike, the bonfire has started up."

I jumped out of bed and ran downstairs and outside. The bales of straw had finally caught and were well ablaze. Speeding up the track came a fire engine. It was only the bonfire that was burning. However, someone in the village had seen the fire in the distance thinking it might be the unit that was in trouble. We had not seen it as none of our bedrooms faced the back of the house.

The firemen still had to extinguish the fire. My wife made them all a cup of tea and they all had a bit of a giggle about our disturbed night's sleep.

The second fire we had on the farm was also manmade. At the far end of the unit was an old muck heap that had been there for years. It had rotted away over the years and was very dry on top. We should really have shifted it before we started building the unit, but we never got round to that.

One day we had some rubbish to burn. Nick said he would sort it out. I told him to take it as far away from the unit as possible but not light the fire anywhere near the muck heap. Unfortunately, Nick did not take my advice. All seemed ok on the day. However, the following day I noticed the muck heap was smouldering away. At first, I thought it was just isolated to that one part of the heap. I took a muck fork and scrambled up onto the top of the muck heap where the smoke was coming from. The smoke was coming from deep in the heap. I forked away the top couple of feet of dry muck to reveal the glowing rotting muck underneath. I dug some more muck away further along the heap to find the same thing. There was that much heat built up it was spreading through the whole heap. I got Nick to come and help. We had to rig up a long hose pipe and started dousing the muck heap. We were on that job for days, forking away and dousing down. Eventually we managed to get on top of it. Another lesson learnt.

The years were flying by. Our son was excelling at school. Our youngest daughter was on her way to owning her own travel agency now and our eldest daughter was married to Sammy's 'best friend'.

Outdoors

There was more expansion in the pipeline. Jim decided to start a new venture. Outdoor pig farming. This method of pig production was becoming very popular, with some pretty large scale farms being set up. It was supposed to be a less expensive alternative to indoor production. It was in actual fact a lot less expensive to set up. You needed some decent free draining land, usually rented. Then all you needed were the sow huts, fencing, water troughs and a tractor or two. Another reason for going outdoors was pressure from the 'do gooders' that reckoned indoor pig production was cruel. As it happens it is the same 'do gooders' that began to complain about the sow huts being a blot on the landscape. One suggestion was that they could be sprayed with pig slurry to blend them into the surroundings. Oh dear!

The decision was made. We would set up three four hundred sow units outdoors. John asked me if I would be happy to oversee all three units. Another big adventure and challenge. How could I refuse?

Rob, my assistant manager on the Crayke unit, would take over as manager there. He would take on the farmhouse too, so we decided to take the plunge and buy a house of our own. By this time both our daughters had left home, so it was just our son and us two to move.

Jim had a personal assistant at this point in time as you know. Basically, he was the overall farm manager, although mainly on the arable side of the business. It fell to him to source the land for the outdoor units.

The first site he found was a unit that was already up and running and was owned by a feed company. Jim bought the whole outfit and kept the two men on that were running the unit. At the time he bought it it was set up on a disused airfield in Holme on Spalding Moor. Fortunately, just across the road from the airfield there were fields belonging to a farmer who was happy to rent the land out. Having a unit that was already set up was useful for me to get an idea of how to plan and lay out the other sites.

The idea of the outdoor system is that you fit in with the crop rotation. So, in theory after two years in one field or fields you move on to another area of the farm. Working on the basis of ten sows per acre of land for a four hundred sow unit you need two areas of forty acres available to use.

The plan was that once the land was found I would set on marking out the paddocks. I did most of this myself. Putting posts in. Laying out the black alkathene water pipes, water troughs and wiring up the paddocks with the electric fencing. A word of warning here. When you look at the massive area you have just found it looks far too big. However, once you start wiring up the paddocks it takes on a Doctor Whos' Tardis in reverse type affect, in so far as the paddocks look too small.

Jim had bought me a Peugeot 205 van and a small trailer to ship stuff around in. I used a supplier local to the main farm in Milby. They had just about everything I needed.

The wooden posts that formed the paddocks were three to four inch diameter and about five feet long. I rented a tractor and post knocker to put most of them in. It was possible to knock them in by hand but not ideal when you have over a hundred to knock in. The posts formed the corners and gate openings of the paddocks. The electric wire was a seven-strand wire. The wires formed the boundaries of the paddocks. Two lengths were placed at about a foot and two feet off the ground and tensioned by hand and secured onto the posts by plastic bobbins. A water trough was placed on the edge of each paddock. The sow huts were mainly the half round ark type made of corrugated sheets fastened onto an angle iron steel base. They had plywood fronts and backs bolted onto an angle iron frame. The front had a hole as a door, the back had a small, hinged flap for ventilation. These huts would hold ten fully grown sows comfortably.

The first farrowing huts we had were a similar design but individual size. A floor area eight feet by four feet. They had a metal fender hooked under the front of them to keep the piglets in. This worked for about the first two weeks until they learnt that they could jump out. We had six to eight huts in each farrowing paddock. The sows were allowed the freedom of the paddock. They always returned to their own huts once they had claimed one to farrow in. The piglets also returned to their own huts when they had got to that adventurous stage.

The layout of the farrowing paddocks was simple, with just a central passageway with rectangular paddocks either side. The gateway into each paddock had a white electrified tape across with a spring-loaded hook handle on the end which fastened to the wire at the other side.

The best way to describe the sow paddock system we used is it was like an aerial view of a sliced cake. The idea being that each slice held one weeks' worth of sows. We were weaning once a week. So however many sows we weaned in any particular week would go in one slice together. The boars we had ran in packs of three and they were introduced to the weaned pen of sows just after weaning. It was just a matter of the boars helping themselves as the sows came on heat. Crudely referred to as a 'gang bang' system. If you didn't actually see a boar serving a sow you could see marks on their backs if they had been ridden. The sow was then spray marked and her tag number recorded.

With this sliced cake system fifteen paddocks were needed, each paddock at a different stage of pregnancy. A pack of boars would run with the sows at the three week stage and six week stage to catch any that may return to heat.

In the centre of the cake where the paddocks narrowed down was a circular area with wooden gates to access each paddock. This area had a passageway running off it to the outer edge of the cake and was used to move boars and sows around.

This system of serving was not the best but was the popular one then.

Later we did try a 'Dynamic group' system whereby you have several large square paddocks with up to 50 sows in each. With this system it was a matter of adding a few sows to each group at weaning and taking sows out that were due to farrow. So, in essence there was every stage of pregnancy in each paddock. The pack of boars would stay in that group to serve and catch returns. This now seems even more haphazard.

One of the drawbacks with any outdoor system is the seasonal differences with weather and daylight hours. Let me elaborate. In summer, when the weather is dry, bedding up the sow huts is not necessary as frequently. In winter bedding was needed on a weekly basis but with less daylight hours to fit all the work in. Added to this you had frozen water pipes to contend with. Even the need to bowser water around the units if there was a prolonged cold frosty spell. You may ask why the water pipes are not buried underground? This would have been too large an undertaking and where the pipes join onto the trough it would still freeze anyway. It was far more practical to let the pipes thaw naturally. There is a tale concerning frozen pipes which I shall tell later.

Within six months we had the three outdoor units up and running. On the first unit the piglets went off to a farm up the road that could house them through to thirty kilos. They were then sold to be grown on elsewhere. On the other two units we decided to house them in weaner huts on site. These huts were a low lean too design with a feed hopper on a plinth inside where there was a hinged lid for access. The outside run had hurdles pinned together giving the pigs an area twenty four feet by eight feet outside and the same area inside. There were around thirty pigs in each hut and were kept on site up to thirty kilos.

Land had to be set aside for these as once the huts were emptied, they were lifted and moved onto fresh ground.

We copied the design of these huts from a firm that was charging a small fortune for them. Chuck and Leo made all of the huts we needed, for a fraction of the cost. As far as I remember we had about twenty on each unit. To move them in the field there was an axle with a big wheel at either end. Once the hurdles had been removed the hut was lifted with a chain and the axle slid into position. At the other end of the hut was a short drawbar. Chuck had made an 'A' frame that fitted onto the three-point linkage on the tractor, which was then backed up to the hut, fastened with a drawbar pin. Then lifted and towed to the fresh piece of land.

By this time the business had three indoor units of four hundred sows and three outdoor units of four hundred sows, along with the heart valve business. Jim in the meantime had also invested in mushroom growing and a nursery producing cut flowers. We were now a LTD company. It is now 1993. Diversification was the way ahead it seemed.

Before I go further into the ins and outs of the outdoor units, I need to point out something that was going to affect every indoor pig producer in the country. In six years time the government was to bring in a sow stall and tether ban. Something associated with being in the EU. Farmers were to be allowed an eleven year phase out period. But many, who realised it was inevitable, went straight ahead with expensive conversions. This was to be a massive blow to producers. It meant a whole new system would have to be adopted on the farms. In a nutshell, buildings had to be altered to enable the sows to be housed on straw in open yards. The easiest and most efficient way to feed them would be with an electronic sow feeding system. (ESF) This all involved enormous unwanted investments. It was to see the demise of many producers. Jim was one of the many who decided to plough on with the alterations sooner rather than waiting for the deadline.

Around this time Jim also started renting an indoor pig unit on the outskirts of Grewelthorpe, a small village near the market town of Masham. This was a very old unit with wooden clad buildings and portacabins for farrowing houses. Some of the buildings needed converting to hold bacon pigs This meant digging slurry pits out and walling them up. Our builder and his mate were employed to do this and somehow, I ended up there helping on a few occasions. This was the late eighties and the unit was very old then. Keep that in mind as I was to return there many years later.

Back to the intricacies of outdoor pig farming.

One of the main advantages of this system is that it is less labour intensive. Add to that the initial setting up costs and it does seem very appealing. Finding the right type of land is of paramount importance, as once you have signed on the dotted line for a couple of years you are stuck. Quite possibly literally, as we found out to our cost.

On the first unit over at Holme on Spalding Moor, feeding was done by hand. The feed was bagged up from a bulk bin and carried into each paddock. This meant striding over the electric fencing wires, wrestling your way through the hungry sows and leaving a trail of food across the driest piece of paddock you could find. This was a really inaccurate way of feeding but, consistently inaccurate. Kieth, the manager on this unit actually drove the tractor and trailer into the paddock with the bags of food on. Then his mate would drop the wire to let the tractor move into the next paddock. They would drive round the 'cake' halfway up the 'slice', so to speak. This did make feeding easier, and the paddock was not as churned up in that area. The drawback was that eventually the wires would slacken off and therefore needed tightening every so often.

We eventually employed a third man on this unit who would in time become the manager when this unit was moved to another site and the current manager moved to a new unit.

The son of the farmer that owned the land we were using on this unit was actually a lad I was at college with. It's a small world in farming. We used straw from his land for all our bedding. It was the best straw I have ever come across. He never rushed into combining or baling his straw. His philosophy was that 'it would be ready when it's ready'.

We had two years on this group of fields. In the meantime I was setting up another site just north of York in Sandhutton. Kieth and his assistant were to run this one. The third man was going to take over running the original unit.

Current music trend…...The Cranberries

Fashion…………................lots of plaid apparently

Chuck and Leo had been busy in the workshop and were now producing our own farrowing huts. We went off the half round ones and moved to a far better design. Mainly made from plywood with an insulated flat steel sheet roof. These still had the same floor area but because the sides were almost vertical it allowed the sows more freedom to move inside. Because of the improved design it reduced the risk of piglets being laid on.

The site I was setting up in Sandhutton was on one large field. Because of the shape of the field, I decided to put the 'training paddocks', which is where all fresh breeding stock coming onto the farm are kept, over in the far corner of the field. Then there was the 'cake' which took up most of the centre of the field. The farrowing paddocks were in one long line running parallel with the lane which led up to the farmyard.

We were going to stock this site from scratch. So I had a tight deadline to work to, with new stock arriving on a prearranged date. It was quite hectic at the time. I was working flat out to get things ready.

The training paddocks for the gilts were only small as they only needed to hold a dozen, weighing about one hundred kilos. The gilts needed to be trained to understand the electric fence. To do this the perimeter of the paddock was fenced with pig netting fastened onto wooden posts spaced about twelve feet apart. Inside the pig netting I put two strands of electric fencing wire. These paddocks had their own power supply to the fence. It was fun off an ordinary twelve volt battery and through an energiser. A freshly charged battery was plenty powerful enough to deter the gilts from trying to escape.

I was carrying out the finishing touches to the first training paddock when I saw the wagon arrive at the far side of the field. I made my way over to let him know he could back all the way across the field to the paddock. He was not convinced, but it was dry, and he did manage it. He backed right up to the entrance of the paddock. We let down the tailgate and ran the gilts into the paddock and they ran around enjoying their new found freedom. I put the tape gates across the gate and some pig netting. One gilt touched the wire, then another. Nothing happened. Then I realised I hadn't flicked the switch on the energiser. I was in the paddock with the gilts. In a panic, I straddled the wire to reach over to flick the switch on the energizer. I heard a gilt squeal as she touched the fence, then another and another. In s rush to get out and sign the paperwork for the delivery so the driver could get away I managed to get tangled in the wire. Ouch ouch ouch, well the fence was definitely working. The gilts soon learn not to touch the fence. It doesn't give them a massive shock, just a tickle but enough to deter them.

Because the sites were normally well away from any buildings it was not possible to get a mains electric supply to the fencing. For this reason we either used a solar panel to keep the battery charged up or a windmill. The type you often see on yachts. Only one site would be powered straight from the mains, but still passing through an energiser which adjusted the voltage. This was the most reliable way to power the fence. Even if a wire had snapped and was shorting out the system it would remain powered up. As opposed to using a battery, once a wire was down and going to earth the battery would drain quickly.

The shorting out problem was also caused if grass and weeds grew and touched the wires. For this reason regular strimming was necessary. The sows were very quick to cotton on to the fact that there was no power in the fence if a battery had gone flat.

We now had three units up and running. Sow food was delivered to two of the units in one ton tote bags. The other unit had access to a bulk bin so the manager there kept on the bag feed system. The two units with the tote bags had a feed trailer each. The trailers had two one ton tote bins attached to the base. This enabled two different rations to be fed, although we were still bucket feeding.

Later as we learned more about outdoor pig producing, we bought a feed thrower for each unit. These worked off the hydraulics of the tractor and the amount of feed could be calibrated to suit the stage of pregnancy and number sows in each paddock. By this time our mill on the main farm at Milby was able to pellet a twenty two millimetre cob which meant less wastage.

The unit at Holme on Spalding Moor was ticking over nicely now as was the unit closer to home in Aldwark. Unfortunately, our stay at the site in Sand Hutton did not last long. The farmer who owned the land decided we were causing too much damage to the soil structure. A misguided opinion as the pigs do more good than harm. So, prematurely, we had to find a new site and move again.

Jim's personal assistant found some land over on the east coast about three miles inland from Filey. I had a ride out to see how I could lay the unit out. It was a large field, on a slight slope. The soil looked pretty sandy but was black. The farmer had the field sown with Linseed which was in flower at the time. He was keen to get pigs onto his land. So keen that he chopped the Linseed down. Apparently, he could get paid more for doing this than if he left it to ripen and harvest properly. Something to do with Government subsidies. It was early summer when I started setting up the unit, it was nice and dry. I built the 'cake' on the brow of the hill, the farrowing paddocks were to one side of the field. I left a large area of land at the bottom of the slope near the entrance to the field for the weaner huts.

As summer neared its end we had managed to move everything off the Sandhutton site.

We bought a caravan for use as an office and bait room. It was a large caravan. We didn't think it would move far in windy weather. However we underestimated the strength of the winds whipping off the east coast. Kieth gave me a ring one morning,

"You best come and look at this." he said.

I arrived to see the caravan on its roof up against a tree about twenty yards from where I had positioned it. It was in quite a mess. They are quite fragile. We managed to stand it back up and do  a bit of panel beating on it, this time we put small bales around it so the wind couldn't get under it. We also roped it down.

I was up at this unit helping them bed up one day in the summer before the weather changed. It was very windy. We used barley straw. If you see a crop of barley in a field it can be identified by its barbed awns at the tip of the seeds. When the barley goes through the combine harvester the seed is separated from the straw and the awns come of the seeds. All this is baled up. Anyone who has worked with barley straw with know how aggressive these pesky awns can be, especially if you get them in your underpants. The odds of what happened to me must have been pretty high. I managed to get one in one of my tear ducts. I have had them in my eye before and been able to wash them out. This one was embedded in my duct. I tried to tease it out between my finger and thumb but there wasn't enough of it protruding. Kieth spotted I was struggling.

"What's up Mike?"

"I've got a bloody barley awn in my eye."

"Let's have a look, Oh nasty, its right in your tear duct!"

Do not try this at home children!

"Give me your penknife Mike."

"Eh? What do you have in mind?"

With that he took his penknife out of his pocket, and I handed mine over. He opened up both blades and told me to keep my head very still. With surgical precision Kieth used the tips of each blade to nip the end of the pesky barley awn and gently ease it out. Not ideal but when you are out in the middle of a field, needs must.

Was the ease with which the caravan was overturned a bad omen? You may say yes. The weather changed dramatically in late September. Rain, rain and more rain. We soon found out that the soil was not sandy at all. It was more like peat. The area where the weaner huts were soon became a bog, waterlogged. Feeding the sows became a challenge to say the least.

We needed to relocate the weaner huts. Remember the axle and hitch method? It was so wet and boggy the four-wheel drive Renault we had on the unit could not cope. We ended up using two of the farmer's four-wheel drive tractors fastened on to each other. It was an absolute nightmare. The pigs had to be loaded into the stock trailer. Then the hut and hurdles. All towed up to higher dry ish ground. Then back down for the next hut. Twenty times.

It was at this point we realised this had been a bad move. No wonder the farmer wanted this field out of his crop rotation.

The weather did not improve. I think it was the wettest winter on record. I went up one weekend to help the manager feed the sows. The icy rain was blowing horizontally straight off the coast. It was horrendous. The ground was so boggy you could hardly walk with a bag of food on your back. Not to mention there was no dry patch to put the food. Fortunately, not all of the sows were hungry enough to brave the weather. I even went to the extremes of sourcing some, ex-coal mine, thick rubber conveyor belt to lay on the ground to put the food on. This was not that successful, long term.

We managed to survive the winter much to the credit of the manager and his assistant. In the spring the decision was made to move the unit yet again. We found some good land in a village called Rillington on the A64. The land was sandy, free draining and flat.

I had two large fields to work with. One I set up for farrowings and weaners, the other for sow paddocks and gilt training paddocks.

This site was near a housing estate. The famous Abraham Lincoln quote, "You can please all of the people some of the time, some of the people all of the time, but you cannot please all of the people all of the time," could well have been associated with pig farming. Due to public pressure lots of producers decided to venture into outdoor pig production. However, as I mentioned earlier it was not without its critics. On the whole the same people that forced the move were the same people that ended up complaining about the units being a blot on the landscape as I mentioned earlier.

On this new site people from the housing estate could access the unit to see close up what a pig looked like. At first they seemed happy with a pig unit in their back garden, so to speak. That is until we started our usual routine of burning off the straw beds when we moved the farrowing huts and weaner huts onto fresh ground. This was a practice advised by vets to ensure any disease was burnt off. We always made sure the wind was blowing away from the housing estate. This was not good enough though and several residents complained directly to the Environmental Health. I was on the unit when one of their officers arrived unexpected and uninvited. I was told in no

90

uncertain terms that we could not burn the beds. I explained it was a pig health issue but that did not wash. So a simple job became a major one in so far as every time we moved any huts, which was on a weekly basis, we had to scrape the beds up with the tractor, load them into a trailer and form a muck heap in the corner of the field.

The next visit I had on this unit was from a chap from English Heritage. Apparently, the field I had set up for the farrowing paddocks was possibly an Iron Age burial site. He showed me an aerial photograph of the field. There were the tell-tale circular shapes on the photograph. He wanted to know how deep I knocked the wooden posts into the ground. And insisted I do no more. Fortunately, the unit was set up by this time.

Meanwhile on the other two units things were running fairly smoothly. With the odd hiccough here and there.

Let's start with the unit at Holme on Spalding Moor. It is early summer, the temperature is warming up so I suggested to the manager we should dig out some wallows for the sows to roll around in to keep cool. I bought some spray attachments that were simply self-tapped into the alkathene water pipe that was laid round the perimeter of the sow paddocks. These could be turned on and off when needed. So shallow holes were dug out on the edge of each paddock and then filled with water. The sows loved it, and made quite a good sloppy mess to cover themselves in. This formed a sun barrier and stopped them burning. Within a few days we had a surprise visitor. I was in the bait room with the manager and his assistant.

"Are we expecting visitors Dave?"

"Nope"

"Well, we have one"

I went outside to greet whoever it was. As the van drew closer, I noticed RSPCA on the side of the van. The lady parked up and got out. Before I could say anything, she spoke.

"Hello, it's ok, I have seen what is going on. Someone called us saying a pig was stuck in the mud and was struggling to get out. I know it is a wallow to keep them cool. But when we get a call, we have to follow it up."

"Ah, well I am sorry you have had your time wasted."

91

Off she went with a smile on her face.

Someone Jim knew had an outdoor unit over in Cheshire and had a similar experience but on a larger scale. Apparently, he was heading home one evening and his route took him past his unit. As he approached he saw blue flashing lights in the distance. His first thought was maybe his straw stack was on fire. When he reached where the fire engine and a police car were parked he saw what was going on. There were three firemen in one of the paddocks trying to put a strap underneath a sow that was happily laid in a wallow. He told them what she was doing, and she was not stuck. The police had had a call saying that there was a sow in distress.

Following the hot summer of this year was a bitterly cold winter. This meant frozen water piper and drinking troughs. Sows will only survive so long without water so it was a matter initially breaking the ice on the water troughs and hoping the pipes would thaw out. One weekend I was in a panic over the water situation. The pipes had been frozen for two days. I needed to get water for them all. We had an old slurry tanker on the farm which was used just for water. I rigged it up onto the tractor and drove to the canal where I could draw some water out. It would take a few trips but I had to get water to them. I got back to the unit, the canal was only a short distance away. I drew up alongside the first paddock and connected the outlet pipe. The ground was rock solid in the paddocks. I broke through the ice on the first trough and tossed the big pieces of ice away. I started filling the first trough, thinking the sows would come running out for a drink. Such is the nature of sows that one or two tottered out, struggling on the hard ground. They looked at this crazy man working out in the freezing cold wind as if to say, "You must be joking, we are not that thirsty."

Eventually winter gave way to spring.

We were on this site for a couple of years and then the landowner wanted his fields back, so we found some land just south of Selby. This was the site where we tried the dynamic group system for the sows. Mainly because of the shape of the fields not lending themselves to the cake shape. I was down on this unit one day when a chap rolled up in his shiny new Jaguar.

"Are you missing a pig?" he asked, "We have one cornered up at the golf course."

"Not that we know of, but we can come and have a look"

Dave, the manager, and I hooked up the stock trailer and followed the chap up to the golf course almost a mile away. When we arrived it turned out it was one of our gilts that had managed to escape from the training paddock and wander all the way up  there. It was quite funny, there were four blokes in their golfing attire trying to hold the gilt in the corner of the car park, waving their golf clubs as if she was about to attack them. She was not at all bothered and was just standing there snuffling around. Dave manoeuvred the tractor and trailer close to her, and I opened the gates on the back. We had brought a bucket of food up with us just in case. I tipped a bit on the floor of the trailer and told her to get in. To the amazement of the golfers she walked straight in as calm as anything. Thank God she didn't get on the eighteenth green. The club could have ended up with an extra bunker. They do tend to dig a little.

Part of the perimeter of this site was bordered by Selby Canal. Not far from the canal was Drax power station. At the time there was a big hoo-ha about possible air pollution from there. I didn't believe it at first, however, on a drizzly day when the wind was blowing in the direction of the unit you could feel the wet tingling your skin. There was definitely some form of acid coming from the cooling towers. Someone confirmed this telling us that the paintwork on some cars in the area suffered as well.

We moved to this site at the beginning of summer. It was supposed to be a really hot one. One day in the middle of harvest we noticed a lot of smoke coming from the other side of the canal. It was a couple of fields up the road from the site. We guessed someone's cut straw had caught fire. The wind was whipping it up. The canal was at least twenty yards wide so surely the fire was contained at that side. The next thing we heard was a fire engine coming along the road at the top of our fields. We went up to see what was going on. Keep in mind that the soil in this area is very,very sandy. The fields also had quite a slope on them from the side of the road going down to the canal. The fire engine pulled up two fields along from where our fields started.

"What's happening?" I asked the driver.

"We are going to drive down this field and spray water at the bottom just in case the fire jumps the canal."

"You best be careful that you don't get stuck, this sand is really soft."

"We will be ok, don't worry."

He proceeded to back into the field and only got about twenty yards in before his wheels started spinning and sinking into the soft sand. At this point we spotted that the fire had jumped across the canal and was moving up the swathes of straw in this field. It was moving at a pace now.
As opposed to the fire engine that was totally bogged down. Despite this the driver stayed at his wheel revving away and digging himself in deeper and deeper. We decided to stand well clear along the road. The driver's mates were shouting at him to jump out. In no time at all the flames were licking around the back of the engine. The driver still didn't get out. His mates ran over and virtually dragged him out coughing and choking in the smoke. They managed to release a water valve on the engine which flooded out around it. They were batting away at the flames with those rubber flap things they have. Fortunately, the wind dropped at this point, and they managed to get the fire under control. The fire engine did not survive, the back tyres burnt out and blew. Luckily the fuel tank did not catch the flames. The fire never got into our fields mainly because there was nothing flammable.

For a small amount of time we needed to employ a couple of men to fill a gap when one young lad left this unit. Someone suggested I should ask at the local jobcentre to see if they had anyone on their books. The first chap they recommended was keen enough but had no experience with livestock and I am sure he must have told the staff at the job centre that he had. He did not last long with us. A lasting memory of him was the sight of him running across a farrowing paddock with an eight kilo pig in his hands, hot on his heels was an overprotective mother. It wasn't even her baby, but the squealing had set her off. We were shouting,

"Drop the flipping piglet," as he tripped over the electric wire.

Why was he carrying a piglet? I hear you ask. We were in the middle of weaning. We used to walk the sows out of the farrowing paddock in the morning of weaning day. After lunch we would go in the paddock with the stock trailer and back up to each hut in turn. Invariably by this time all the piglets were back in their huts, so it was just a matter of running them out onto the trailer. Occasionally an odd piglet or two would run into the paddock next door looking for their mum. Hence the one the chap had caught and was running with.

The second chap I took on from the jobcentre was marginally more useful. In fact, he knew everything, apparently. He used to scurry around at speed like a hamster on drugs. I seem to remember he managed to dislodge a wall in the entrance to a barn with the tractor but denied all knowledge of it. He did not stay with us for long.

The third site that was now up and running was the one near the main farm at Milby. This one was up a sandy track in the village of Aldwark. Two brothers ran this site for a while until they both left around about the same time. While they were there, they did a good job and made this probably the tidiest of the three sites. There was a large acreage of sandy land in this area which meant after the first two years we could just move across the track and start again if and when necessary. I set this site up with the 'cake' design again.

We had Duroc boars on this site in groups of three. Duroc boars are as 'hard as nails' and very temperamental. For no reason at all they would suddenly fall out with each other and start having one almighty scrap. There was no way you could separate them by hand. They would quite easily fight to the death if left. They would snap wires as they fought going round and round head to head. The only way we had to stop them was to drive the tractor between them. On one such occasion one of them was left with a cut along its rib cage about eighteen inches long and deep enough to see his ribs. Just a flesh wound as far as this Duroc was concerned.

Another day I visited the unit when the manager was supervising some serving that was going on. One of the Duroc boars was merrily serving a sow. He was backed up against the electric fence wire with his testicles directly on it.

"Your fence is not working."

"It is, I tested it earlier" said the manager, "Try it."

I took a blade of grass and touched the wire. This is a tried and tested method. I moved the length of the blade of grass up the wire. I definitely got a shock. As I say, the Durocs are as 'hard as nails'

After two years on this field we moved over the track onto some fresh land. This land was actually owned by a different farmer. A short time after we moved the farmer who owned the land we had been on decided to put pigs on it himself. Maybe not such a clever idea from a pig health perspective, but there was nothing we could do about it.

This new site was just as sandy as the fields across the track. The feed in the one ton tote bags was stored on wooden pallets at the top of the field. This was the sandiest part of the field too as when it was windy the sand was blown up to this end up against the hedge.

One wagon driver arrived once with his twenty ton load and decided he could get in and out of the field easily. I did offer to unload him on the track, but he was determined to come as close as possible to where we stored the feed. He tried to turn his wagon around in the field before I unloaded him. Needless to say he jack-knifed and managed to bury his driving wheels in the sand. I unloaded him where he was then towed him out with the tractor.

Shortly after the move the two brothers who were running the unit left. One went to manage an indoor unit and the other went back to building work, I think. So, I took over the running of this unit and found another chap to come and help. He was very eager to please and wanted to try and do everything for me. His eagerness almost led to what could have been a fatal accident.

We needed to fill the feed trailer up with food. To do this we used the tractor with the front end loader with pallet tines on. On the top of the tote bags there were four loops of webbing that had to be fed over the pallet tines. It could be a one man job but is easier with two. I was on the tractor, the new chap was standing next to the tote bag. As I drove in to locate the tines in the front two loops the new chap for some reason managed to position himself between one of the tines and the tote bag. I put my foot on the brakes and stopped the tractor as the pallet tine started to push against his chest. He was bruised but could quite easily have been punctured. Such a lucky escape. It just goes to show how easily accidents on farms can happen.

Piglets on this particular site used to go off at weaning time being sold to another company. This meant we only had the sows and litters to look after.

The recording system we used was the same as we had used on the indoor units. It was a PIC Pigtail system. It was quite foolproof. There were small recording cards the size of a pocket diary. Any event that happened was recorded on a card. One card per day. Farrowings, fosterings and deaths were recorded on one set of cards kept in one folder and services and returns were recorded in another folder. These cards then went back to our main office to be updated on the main system by our secretary. When the sows were weaned and served again we would get a new main record card for each sow with all her details on and previous performance figures. For example, piglets born alive, piglets born dead, piglet deaths and the cause, piglets weaned and what parity she was.

In the bait room, which actually was a forty foot wagon container, I had a large pin board on the wall with all the paddocks mapped out on it. Each sow had a large notice board pin with her number on it. Whenever the sows were moved, the pins were moved accordingly. All the sow record cards were kept in a cabinet, sectioned off into the corresponding paddocks too. This made life so much easier.

The tote bags that the sow food was delivered in could not be re-used to supply other pig farms due to the risk of spreading any possible diseases. Normally they were sent off for recycling. They were a woven plastic type material. One day, in my infinite wisdom, I decided I would burn a whole bundle of them. This turned out to be a really bad idea. At first there was just a little bit of smoke. Then they suddenly caught, and a massive black plume of smoke started billowing into the sky. It resembled the plume you might get from a nuclear explosion. All  the sows were totally spooked by the smoke and set off running in a blind panic, breaking through the wires that separated the paddocks. Just about every wire was snapped and the sows all ended up down at the bottom of the field. As if it could not get any worse it also spooked the sows at the other side of the track belonging to our previous landlord. They broke several of his wires too. We spent the next few days rewiring and sorting out which sow belonged where.

It is now 1999. Money had been ploughed into the outdoor units, the indoor units, the mushroom farm and the flower nursery. Dark clouds were about to gather over the company though in the next few months. Whatever the series of events that followed or who was ultimately responsible for what was to happen, only a few will know.

I was working on the unit at Rillington when I got the news that the business had gone into administration. The office at the main farm was taken over. Nothing could be bought without it being authorised by the administrators. Obviously, we still needed drugs etc on the units. Not even a syringe or box of needles could be bought without it being authorised.

All the firms and suppliers that we had been dealing with must have been informed of the situation immediately. All payments were frozen.

I was on the far side of the unit when the unit manager rang me.

"You best get over here Mike, Thomas maintenances are here and are going to take the wheel bearings off the stock trailer and tractor that they put on two weeks ago because they haven't been paid for."

I jumped in my van and rushed over.

"What are you doing?" I asked.

"We are taking everything that hasn't been paid for yet."

The manager was about to start weaning and needed both the tractor and trailer.

"Is there a solution?" I asked the mechanic.

"Pay us now and we will go."

I got straight on the phone to the office and explained the situation. Jim's personnel assistant was issued with a cheque from the administrators and drove straight over. An eighty mile round trip. Stress levels were high. Jim must have been distraught. This was just the start of it though.

Later that week I visited the unit down at Selby. I was told that the firm we leased the tractor from were coming to take it away. There was nothing we could do to stop them. Payments had stopped and the tractor was theirs as it was leased. That afternoon they collected the tractor.

We were helpless without a tractor so I decided to ask our landlord if we could borrow one of his temporarily. Fortunately, he said a reluctant yes.

Things were happening so fast, and we were battling to keep all three units running. Eventually the inevitable happened. Money had to be generated somehow. Assets needed to be sold off. The three outdoor units were bought by a large feed firm. The main farm was bought by another local pig producer. The farm at Crayke was also sold off to another local pig producer.

The feed firm brought in their own Production manager. I was on the unit at Selby when I first encountered him and his assistant. It was a Thursday. Weaning day. Dave and I had started early to get on with weaning before they arrived. They pulled up in the Production manager's VW Golf GTI. They jumped out, opened the boot and took out a roll of that orange plastic stuff that you see around minor road works. They came strutting down to where we were.

"Right, we are going to show you how to get sows out of the paddock!" the loudmouth manager said as they unrolled the orange stuff that they had pinched from the side of the road somewhere.

"It's ok. You don't need that." I told him. "They just walk out when we put a bit of food down."

The sows were already spooked at the sight of the orange.

"Well, this is how we are going to do it," he said, arrogantly.

I'm sorry to say that at this point I flipped. I had never felt like this in my life. I told him in no uncertain terms to stuff his job, he could do whatever he wanted. I left them to it, got in my van and drove. I headed for home but parked up in a layby wondering whether to go to Milby to let Jim know what had happened, feeling as though I had let him down. I decided to ring him instead. I managed to get hold of him and explained what had happened and how I was feeling. He had no answer. I had never heard him sounding like he did. He sounded so tired and defeated by everything that was happening. He simply told me that I must do what I thought best.

At this point I felt we were very near the end of what we had built up together over the last twenty five years. I cannot imagine how he must have felt. He had started from nothing with the pig side of the farm and built it up over the years.

The feed company that had taken over the outdoor units did call me to their offices to ask me if we could work around our differences. I suggested they get rid of their Production manager. They would not and did not. I did not go back to work for them. I was signed of sick with what the doctor reckoned was depression. In time, I started looking for a job elsewhere. To this day I do not know what the future held for the outdoor units. I know the unit at Crayke is still operating with Rob as manager. I also know that the unit at Milby is still operational. The original buildings are still standing. I drive past them at least once a week as we now live close to the farm again.

Current music trend…Westlife and Britney Spears

Current fashion trend……..vest tops and leather jackets

Suffolk

We became somewhat Nomadic over the next few years. I found a management job on a large family run business in Norton, Suffolk. The business had been run for many years by the head of the family. He had now, supposedly, retired.

His eldest son had taken over managing the business, with his brother and his nephew taking on different sections of the enterprise. They had a business manager, a line manager and two permanent office staff. There was also a team of builders and a fully equipped workshop with several staff. They had over a thousand acres of arable land, their own feed mill and four indoor pig units. One of the units was producing the breeding gilts for the other units. So, the whole business was pretty much self-contained.

I felt as though we needed to make this move to get away from what had happened. The interview was conducted by the oldest brother who was now the main man, his father, brother and the business manager. They seemed impressed with my CV and past extensive experience. Steve, the business manager asked,

"What do you see yourself doing in five years time?"

My answer was

"I will be doing your job,"

"Oh, I don't think that will happen," he said confidently.

Despite my cheek I was offered the job.

The pig unit was at the end of a lane just outside the village of Norton. It was a dead end. The farmhouse we were offered was virtually on the unit. It was a large detached four bedroom house with two garages, a conservatory and a large garden. It needed some work doing to it before we moved in. The windows were the old single gaze metal frames. I insisted double glazing was put in as part of the deal to move. It needed a whole new kitchen as the retiring farm foreman who was moving out had taken all the old units out, believe it or not, leaving just the kitchen sink. The company offered to have the resident builders put the kitchen in for us, but I had seen how steady they worked.

So, I decided to install it myself with the company paying for it. We re-carpeted the whole house, and I decorated throughout over a period of time. We paid for a removal firm to make life easier. I drove down on the day of our move in my Triumph Spitfire. My wife followed in our Ford Ka.

Our son had just started university in Leeds. We still owned the house we had bought when I started on the outdoor units. We decided to sell. In hindsight we should have rented it out. But it seemed the correct decision at the time.

And so, a new life and yet another new challenge. My wife found a job at the doctors in the next village. I was welcomed at the new unit as if I was some kind of saviour. We soon settled into our new life.

All the buildings on the unit were built of house bricks. If you climbed up to the top of the main bulk feed bin and got an aerial view it resembled a concentration camp because of the way it was laid out. There were seven hundred breeding sows on the unit. We kept all the progeny and reared them for bacon. All the pigs apart from the newly weaned pigs were on a wet feed system.

I learnt that the previous manager had run the unit with his son. Apparently neither of them ventured out of the unit's office very often. Neither of them were liked. The manager had been very autocratic, bordering on dictatorial, so I was to learn. His method was to send the men off to do a set job telling them to come back once that job was finished, to be told what to do next. Because of this draconian method of handling his men morale was very low when I took over. Production was poor as a result of this.

I explained on the first morning that things were going to change and that we were a team of seven, the seventh being me.

"I will do any job you guys do and I don't expect you to do anything I wouldn't do. Together we can make this unit work to its full potential. If there is some practice on the unit that you are not happy with you tell me and together, we can sort it."

The lads warmed to me, I think. I could almost see a sense of relief on their faces. Their talents had been so suppressed under the old manager. Part of the problem was that he had been there from day one when the family had set up in pigs. The family did not know much about pigs. For this reason, they assumed the way the old manager ran the unit was the correct way and never questioned it or were afraid to!

Changes needed to be made and quickly. I eased myself into my new team. It was the most men I had ever managed on one unit. On the first morning I was spotted by with a sweeping brush in my hand.

"You shouldn't be doing that." he said.

"Gavin, I am part of the team, whatever you do, I do too." Without a word of a lie his response was,

"You are like some kind of God." I was flattered but that comparison was slightly over the top.

In the first few days of being on the unit I soon found out how archaic the working practices on the unit were. I was in my element. There were so many challenges to overcome and modern practices to put in place.

The unit itself was actually laid out in a very logical way. At one end there was a large barn for straw storage and muck storage from the sow houses and grower sheds. There was another large barn at the side of the unit purely used for straw storage. At the end where the first barn was there were two service houses, flanked either side by a sow house for the pregnant sows. Either side of the sow houses were two long farrowing houses, a bait room and an office.

In the next block there were eight flat deck rooms and seven rows of 'lean too' sheds with ten pens in each row which held the growers from thirty kilos. At the sow house end of this block was the feed mixer room that supplied the sows and farrowing houses. At the other end was the mixer room for the growers.

Then there was a row of three fattening sheds with forty pens in each and a building with about twenty pens for miscellaneous pigs, lame or ill. The fattening sheds had their own feed mixer. Beyond the fattening sheds there were two gilt sheds and another large muck storage pad. Finally, there was the slurry store and lagoon.

It was late May when I started on the unit. Summer was on its way, and Suffolk was warming up. I needed to see how the lads in each section worked and what they thought about the way they had been told to do the job. On the first morning I caught up with Peter, who was to be my assistant manager. He was chiefly responsible for the sow houses. He was in the service house. All the doors were shut. There were no windows on the sides of the building. The only ventilation was a narrow opening in the eves. I opened the central passage door and was almost knocked over by the strong smell of ammonia.

"Oh my God Peter, let's have some doors open and get some air moving."

"We were told to leave them shut as the sows come on heat better," was his reply.

I pointed out what a load of crap that was and we proceeded to open all the doors. Later that day I rang one of the brothers who jointly owned the farm, he was in charge of farm maintenance and the full time gang of three builders and told him we needed windows cutting into the brickwork to help with the ventilation, or lack of it. He came to the unit later that day and took some convincing that the sows would perform better breathing oxygen rather than ammonia. The previous manager had obviously misled and misguided everyone.

The windows were put in over the next few weeks. This was the first of many changes I managed to convince the owners to implement. Any changes I wanted to implement seemed common sense to me. Now I think about it, this was the first unit I had taken on that had a readymade team. In actual fact the previous manager had made the task of me slipping into the role a lot easier than it could have been. Apparently, I was 'a breath of fresh air'. The thing is a good manager is only as good as his team.

There were so many things that needed tweaking and changing, many were just small things that, like I say, that seemed common sense to me.

Peter seemed to have things pretty well sussed in the service area. He had a really calm approach to his work and was very well organised. Once we had opened up the doors and installed the windows his department was sorted. I was confident he was going to prove to be more than capable of being my number two.

In the farrowing houses there were three rooms of twelve crates in each building. The pens were half slatted and sat over a deep slurry pit. The lad that was in charge of this area seemed to have been the most 'downtrodden' by the previous manager. He used to chain smoke 'roll ups'. He also had been allowed, for who knows how long, to drink cans of lager on his lunch break. I couldn't believe it when he pulled a four pack out of his rucksack that first day as we sat down in the bait room.

I managed to suggest it was not such a good idea, at least from a health and safety point of view. Once I had won him over, he agreed and stopped. At least on the unit that is.

One of the main things that needed changing in the farrowing house department was the five procedures they performed on the baby piglets on the day they were born. This was a change that I had to run past my line manager. He was basically a 'Gofer' for all the pig units but was answerable to the business manager. So, my question was.

"Why are we doing so much to the piglets on the first day?"

They were iron injecting, teething and tailing, giving them an oral dose of an antibiotic to prevent scour and putting a five digit tattoo in one of their ears, for identification in the future. The poor things were floored by this and laid out looking like near death from the stress. My suggestion was to split these procedures over day one and day three. Of course, this meant handling the pigs twice, but it would and did greatly reduce their stress. The change was agreed. It seemed logical to me but had never been considered. In fact, we stopped the tattooing which was totally unnecessary.

I would have liked to have made more changes in the farrowing houses but with the minor tweaks we made in the service area and this area the number of piglets being born and subsequently reared increased. So, when I suggested structural improvements in the farrowing houses to our business manager, his reply was

"We cannot justify such changes now you and your team have increased production with things as they are."

I had no answer to that. There were other areas to focus on where improvements would save the business money and make life easier for my team.

The flat decks, where the piglets went at weaning had a heating system I had not seen before. The interior design of each of the eight rooms was pretty standard with a central feed passage and pens either side raised up above a slurry pit. A small storage room separated the two rooms in each section. In this room was an oil fired boiler which in theory heated a bank of galvanised steel water pipes that were suspended above the pens on either side. There was a fan above the door into the room. This drew air in from outside and blew it through a long plastic tunnel which was suspended above the central passage. The tunnel had holes in the sides. The idea being that the heat in the pipes was blown over the pens of piglets. There was a Farm X control box for each room so you could set a temperature curve in order to automatically reduce the temperature and increase air flow as the piglets grew.

This system had been in place for many years. The pipes were corroding and would often leak which in turn caused airlocks in the system and the boiler would cut out. The plastic tunnels were a third full of crud because the fans were not sucking clean air in. I don't think they had ever been taken down and cleaned out. We persevered with this for a few months trying to make it work. Short of ripping the whole lot out and starting again there was not a lot we could do to make it work more efficiently. Eventually I convinced the company to put up a brand new building for weaning the piglets into. This was a big undertaking and a move away from the brick built sheds. We decided on a design that is still used to this day. It was one long building. There were five separate rooms which sat over their own shallow slurry pits. Each room had a small storage area where bags of food could be put. The heat and ventilation controls were in here too. In each room there was a central passage with four large pens on either side which held about forty pigs in each. In the far end wall there were two fans which drew air through the room. In the storage area two four kilowatt electric bar heaters were set into the wall. These rooms were so well insulated being made from a sandwich of polystyrene type material, that we would set the temperature to thirty two degrees at weaning to start the piglets off and by day five the heaters would rarely come on as the pigs produced their own heat.

The whole interior of the rooms was white plastic, walls and ceilings. Fluorescent lights finished the job off. The contrast between this and the old buildings was so great. The pigs performed better and Dan, the lad who ran this department was much happier in his work.

I cannot remember what happened to the old flat decks. I think they were pressure washed and left.

As I worked my way around the unit, assessing the good points and the bad points I soon realised that the general attitude was, 'Well we have always done it like this'. I put this down to the fact that no one had dared to question the previous manager. Not even his employers.

Each department had its own large mixer tank and pump which forced the wet feed around. In the farrowing house there was a tap and a pipe coming off the main line in each room. In the sow houses there was a tap and a pipe above each pen.

I will come back to feeding in these areas later.

Now we move on to the grower section. These were the 'lean too' buildings. Each pen held about twenty five pigs from twenty kilos up to fifty kilos when they would then move to the fattening houses. Later at this stage they were overcrowded.

The gates holding the pigs in were made from chipboard, two gates per pen with a central post that located into a hole in the concrete and a slot in the roof frame at the top. A steel rod that was slotted in place diagonally across each gate gave them extra strength. This was definitely needed as the pigs grew and the level of the muck got higher. These pens were straw based and needed bedding up twice a week. Routinely the muck would need to be taken out from the front of the pens to prevent the gates bursting open. This was a two man job as the pigs needed to be held in at the back of the pen. Bedding these pens up was the only job on the unit that needed a team of us.

Now we come to the reason they needed bedding up so often. Down one side of each pen was a glazed feed trough. The feed line had a galvanised pipe fastened onto the wall which was then secured directly above the centre of the feed trough. The pipe had an outlet on either side. There was a small gate at the side of each pen at the same height as the top of the trough so you could get in to turn the tap to allow feed to be pumped in. Once the feed was mixed in the tank and the pump was started the pigs knew it was feeding time and would stand waiting at the bottom of the pipe. As soon as Gavin, the lad in this department, opened the tap and feed started flowing, all the pigs were fighting to get to the pipe. This caused a lot of the feed to be spilt out of the trough onto the bedding. Not only wasting a lot of feed but unnecessarily soaking the sleeping area. Hence the need to muck the fronts out mid-way through the period of time the pigs were in there.

There was no accuracy in the feeding system. It was all guess work as to how long you left the tap open. It also depended on the consistency of the mix. This was all guess work too.

Eventually I managed to convince the company to experiment with dry feed hoppers in a couple of the rows of grower pens. This meant drinkers had to be fitted in each pen too. By doing this the pigs moved to ad lib dry feeding rather than being fed twice a day. We greatly reduced the amount of straw being used and no longer needed to do the extra muck out.

There was no lighting in these buildings. It seemed to be a common theme running through the whole unit. It is common knowledge now that pigs perform better with decent light for a certain number of hours each day.

The fattening houses were my biggest concern. This was also on a wet feeding system. With the same problem as the growers when feed came down the pipes. The problem in this building was that the laying area was solid concrete with a slatted area at the far end of each pen. The feed splashing all over the concrete caused the pigs to muck everywhere. So, they were pretty dirty all of the time. The whole area was pretty squalid. The lighting in these buildings did not help. In the whole length of each building, hanging from the roof there were twelve one-hundred-watt light bulbs, just ordinary domestic bulbs. There were no windows in these buildings. Within a couple of months our 'in house' electrician fitted fluorescent tube lights in all the fattening sheds. I also pushed for dry feed hoppers in here, but it was going to be too expensive even though my experiment with hoppers in a few pens proved to be a success. The new lighting made a big difference though. At least we could see what we were doing after they were fitted.

The gilt sheds had their own feed mixer tank. This area seemed to be working quite well. Not a lot needed improving here. However, an extra gilt shed was put up while I was there.

Now back to the wet feed systems. These were probably the least accurate way of feeding pigs I have ever seen. Not that the pigs were in poor condition. It was just that there was too much guesswork involved.

The sow houses and farrowing house shared one mixer tank. The tank itself was about twelve feet deep and eight feet in diameter with a mixing propeller in the bottom on a vertical shaft. The feed bin that held the food for this mixer was above the hut that the mixer was in. A shoot with a slide at the bottom of it came through the wall and was positioned over the tank. Water came in via a three inch pipe which led all the way back to a massive water tank above the fattening sheds. The mixing method was simple. Open the tap and start letting the water in. Start the mixer. Then pull the slide on the shoot to drop the meal into the mixer. This is where the guesswork and human element comes in. It was a matter of judging by eye the consistency of the mix. If it was made too thick there was a risk of bunging the feed line up. If it was too thin then obviously the pigs would not get the correct amount of meal. Peter, in the service department, liked a medium thickness which looked about right. Phil, in the farrowing houses, persisted on making a thin mix. At the other end of the scale, Dan, who stepped in for Peter and Phil if they were on holiday, made the mix so thick that it was always in danger of bunging the pipeline up.

Once the mix was at the desired consistency the pump would be started up and it would circulate around all of the farrowing and sow houses. Once circulating it was a matter of dropping the mix into each trough via the flexible pipe above each pen with a tap on the end. There was no science to this part of the process either. When I questioned how much each sow received, the answer was that 'we just fill the troughs up'.

There was no way to calibrate how much water or meal went into each mix.

The grower pigs had exactly the same mixer system. Pure guesswork. These pigs were fed twice a day. The same method, 'fill the troughs up'.

And now, the mystery that was the fattening house feed system. This was the most technical one on the farm. Nigel was in charge of this department. He had the control panel set up how he had been told years before. I obviously wanted to understand how it worked. I had watched it in action but was struggling with how we knew how much feed each pig was getting.

"So, Nigel, how do we know how much feed each pig is getting?" I asked.

"It's three clicks for each pig, simple." said Nigel.

"Ah, ok. So how much food is in a click?"

"I don't know. It's three clicks."

"So, it has never been calibrated?"

"Not while I have been here, and that's four years."

I needed to look further into this. The control panel was not computerised. When Nigel opened up the front panel the inside resembled an old telephone exchange control with dials and cogs and lots of wires. Each pen in the fattening houses had its own corresponding adjuster. If a pen had twenty pigs in it that adjuster would be moved on to the number sixty. Hence the three clicks per pig idea. This system did not have a mixer tank as such. The water and meal dropped into a hopper above a small tank which fed directly into the pump. The part of the system that was associated with the three clicks was in this hopper. Above each pen on the feed pipeline was a rubber diaphragm, inside a round alloy block, which opened and closed with compressed air controlled by an electronic solenoid. I would imagine when this feed system was put in it was cutting edge technology. It was quite amazing, but very old now.

We did end up calibrating it, which improved things. However, there remained ongoing problems with solenoids becoming faulty and the rubber diaphragms wearing. Another problem that occasionally occurred was the water supply would stop. There was no fail safe to detect this. The pump would carry on working and push dry meal only through the pipeline until it could pump it no further and the pump would cut out. If this happened, we had to undo joints in the pipeline and hopefully clear the blockage with the pressure washer.

Current music trend ....Dry your eyes mate

Current fashion trend.........shabby chic

That's enough about the workings of the unit. I am sure some of you will have fallen asleep. Over to incidents and accidents, of which there were a few over the next four years. Most of which were man-made I am sorry to say.

One day we were moving sows into the farrowing house. We normally had all the main escape routes blocked. One gilt managed to squeeze down a narrow passage between the flat decks and a grower shed row. This passage was actually only just wide enough for a pig to fit through. The thing with pigs is, if you give them a wide gap to go through, they refuse. But they always manage to squeeze through small gaps where you do not want them.

Along this passage there were metal poles carrying a main electricity cable. The gilt was squeezing past them until it got about half down the passage. It suddenly dropped to the ground as if it had been poleaxed. I ran down to see how it was. It was dead. Its chest was purple as if it had had a heart attack.

I looked up at the top of the pole. Instead of the wires running through the ceramic bobbins as they were on the other poles, they had been wrapped around the rod that the bobbin should be on. How long the bobbin had been missing was anybody's guess. The outer casing of the wire had worn and had made the pole live! We never found out who was responsible for this, but we had our suspicions. There had been a historic disregard for health and safety. We think it might have been the same person that replaced blown fuses in the plug on the welder with a nail!

The gilt sheds down at the far end of the farm had their own mixer tank for the wet feed. Similar to the sow house the feed bin was above the mixer room. I went in one morning to find the tank completely full of dry meal. Someone had not closed the slide fully the day before when the bin had become empty. A delivery had been made that evening. All three tons of food had gone straight through the bin into the tank. But there was some of the previous days mix in the tank. We needed to dig this out. So, Dan and I set on to tackle it. It took some doing but we soon got all of the dry meal out which could be blown back into the bin at some point. Meal and water starts fermenting pretty quickly overnight. It was beginning to smell. I reckoned we were pretty much on top of the job and sent Dan off to get on with his own work while I attempted to empty the rest of the tank. I put the ladder back down into the tank and climbed down with a bucket and shovel. Right down in the bottom of the tank the fermenting gas was building up to the point of being overpowering. It was reminiscent of the Propcorn incident years before. I struggled to get back up the ladder gasping for air. When I told my line manager about it, he said what they normally do when this happens is bring the large compressor up, which was just like one they use on roadworks to run a concrete breaker. So that is what he did. But once again a slight risk, breathing in compressed air was not really recommended.

Harvest was fun. Both the barns had to be filled to capacity. I usually drove the JCB load all, stacking the straw. The amusing and slightly worrying thing was that the pigmen who had been 'confined to barracks' all year were allowed out on the roads of Suffolk on tractors that had also only been used on the pig units. There was the Ford 3000 that had only been used for pulling the water tank around that was used for the pressure washer. The Ford 5000 had only been used for pulling the muck trailer. Then there was the Massey Ferguson 135. The straw trailers were big too. They would hold forty of the big rectangular bales. Probably weighing at least eight tons in total. Straw was being brought in from fields all over the surrounding countryside. At times there could be seven or eight trailers bringing the straw in as each unit was involved for this job. Stacking the straw was non-stop. On the front of the loader there was an attachment that squeezed the bales as you drove into them to pick them up. Once I got into a rhythm, I could keep the trailers moving.

Gavin was crazy, he always chose the Massey Furguson. It had no cab, so he was open to the elements. This tractor was really far too small for the size of trailer. I remember one evening the heavens opened up just as we were finishing for the night. Gavin was the last one to get back to the farm. He came tearing into the farm absolutely drenched, with a massive smile on his face. When he saw me, his face lit up even more.

"That was amazing," he shouted. He was on a real high. He was into his extreme sports and these conditions were right up his street. I tipped his trailer.

"Shall I go back for some more?"

"No Gavin. The fields are too wet now, we have finished for tonight."

He looked really disappointed.

There was a particularly sharp bend on the road between the farm and the village of Norton that some of the pigmen struggled to negotiate with the long trailers. There was a low stone wall on the outside of the garden to the bungalow on the corner. Without fail, every year the wall got knocked down. Every year the main farm office received the call from the owners of the bungalow. I think our resident builders were on standby for the re-building of the wall once harvest was over each year.

Some of the streets in the villages that the tractors were bringing the straw through were so narrow it is a wonder they didn't leave a trail of broken car mirrors. They did leave a trail of straw. If you can imagine an old Western film where the gunman comes into town, and everyone goes indoors and shuts their doors and windows. Well, this was not Billy the Kid riding into town, it was the Browns harvest gang.

The decision was made to demolish the old building that was used for the miscellaneous pigs. It was rather a labour intensive set up and was redundant really. There was a feed bin at one end of the building which needed removing. I decided to tackle this on my own. It would be a pretty simple operation. Just four bolts holding the four legs of the square bin on the concrete base. I got them undone and decided to gently lower the bin carefully backwards using the JCB with the pallet tines on. I was all set to start when the head of the Browns empire pulled up in his trusty Volvo estate. He was supposed to be retired but could not stay away from the various parts of the farm.

"What are you doing Mike? I will give you a hand."

"It's ok, I was just going to ease it down with the JCB."

"No, no, that will not work," He opened the boot of his car and pulled out a length of rope.

"Right, you lift me up to the top of the bin so I can lash this rope around it".

"Are you sure?"

"Yes, it will be fine."

With that he stepped onto the pallet tines. Now just for those who have no idea what these are. There are two, solid steel about four inches wide and five feet long. They attach to the frame on the extending boom of the JCB. With one foot on each tine and the rope over his shoulder he shouted

"Ok Mike, lift me up."

I was not happy about this but started lifting. So, there I am with my employer's retired seventy six year old father twenty feet up in the air straddling two pallet tines. He was now at the top of the bin. With that he reached over and grabbed the handle of the lid on the top of the bin and pulled himself onto it so he could lash the rope around it. He then eased himself back onto the pallet tines and secured the rope onto the frame on the front of the JCB.

"Ok, Mike, lower me down and I will slacken the rope off as we go."

Once he was back down on terra firma I eased backwards as he kept the rope tight. The bin started to tip over and rested against the tines. I gently lowered it to the ground. His plan had worked but was so risky. He was renowned for being a little gung ho though.

Another story I was told about him involved a tractor and a slurry lagoon. Apparently, the builders had been using a tractor for something on one of the other pig units. They had been working very close to one of the slurry lagoons. Somehow the tractor had rolled down the sloping side of the lagoon and was almost totally submersed. Right on cue, our daring pensioner had turned up to find the builders stood looking at the stricken tractor, scratching their heads. With that he sprang into action. Telling one of the builders to fetch another tractor and a chain. Now whether this next part is true I am not sure, but it makes for a good story, and I could quite easily believe it. He proceeded to strip down to hit 'Y' fronts but keeping his flat cap on. He grabbed the chain from the builder and slid down the slope into the slurry lagoon. He then apparently ducked down under the slurry with just the top of his head and his flat cap showing. He secured the chain round the front axle of the tractor and resurfaced to scramble back up the slope. The tractor was then pulled out. I believe this story to be true and you would if you had known him.

One other thing involving our intrepid pensioner was that he hated anything going to waste. He was greatly involved with the demolition of the building mentioned earlier. Most of the timbers were rotten and most of them were nailed together. He insisted on salvaging parts of the rotten timbers by cutting the good bits off. He also had the farm workshop team straightening out the rusty bent nails so they could be used again. It is all true.

Now back to one of my mistakes. In one of the gilt sheds the main upright supports for the roof were round steel posts about a foot in diameter. I thought we would tidy them up as they were rusty and what paint was on them was peeling off. I got hold of some black bitumastic type paint. As we started painting, I realised the paint would go further and would be easier to apply if I thinned it down. What did I have at hand? Diesel. Why not, I thought. This turned out to be a really bad idea. It never dried. I would imagine to this day, twenty years later it is still not dry.

Some of the men who worked on the main farm were not that well travelled. One day one of them came to the unit to sort out a broken feed pump. It turned out that it needed a new part. He rang the farms main office to find out who stocked the parts and where they were based. He found out that they were based on an industrial estate on the outskirts of Ipswich. Ipswich was just thirteen miles up the A14. Most people in the area around us would go shopping there or Bury St Edmunds. The lad from the workshop looked panic stricken.

"How do I get to Ipswich?" He asked.

I thought he was joking, but it transpired that he had never been out of the village he lived in which was close to the main farm.

That reminds me actually of when my wife and I came down the first time for my interview. We didn't have Sat nav and were relying on a road map. We knew we had to turn off the A14 about six miles east of Bury St Edmunds. We took the correct exit and headed through a couple of small villages. We were soon lost. I decided to stop and ask for directions in a small shop.

"Could you point me in the direction of Bacton please?"

All three people in the shop seemed puzzled and huddled together whispering. Then one came up with an answer.

"Bacton? Mmmm, Bacton, not sure any of us have heard of that place." The other two shaking their heads in agreement.

"We reckon you are best going back onto the A14 and trying another exit," said one of the others. I thanked them for their help and got back in the car.

"Wow, that was weird, I'm sure we can't be that far away." I said to my wife.

We had another look at the map and the name of the village we were in was Wetherden. We were actually only three miles south of Bacton as it turned out. It felt surreal, three people who had  not heard of a village that was only three miles away. Maybe they were related to the lad from the workshop.

The company was trying to move forward with the overall management of its pig units and the farm in general. They decided to target health and safety, mainly on the pig units as it was almost non-existent. This was to prove to be a 'can of worms' they may have regretted opening. It was decided that initially the managers and assistant managers would have regular meetings with upper management and the brothers who owned the business. The idea was that we would put forward our and our teams concerns about risk areas on our units.

An accident report book would be introduced at the main farm office. Any accidents of any description would have to be reported.

I seem to remember the longest discussion ended up being over the provision of steel toe cap working boots, and who needed them in which areas. It was decided that the main area they were needed was in the service area. This was because the pens, that the serving of the sows was performed in, were pretty

116

small. Whoever was assisting the serving had to be in the pen with the boar and the sow. So, there was a risk of being stood on. On my unit Peter was the main man for serving but Dan and I also stood in for him occasionally. That meant three of us were identified as needing the steel toe capped boots. However, the boots were only needed really when actually in the service pens. Not that they couldn't be worn elsewhere, the problem was that anything other than rubber wellingtons notoriously rot with prolonged exposure to pig muck. The other question that arose about the boots was that in practice if a sow or boar does stand on your foot it is generally not your toes but any other part of your foot, and so the debate went on.

Another issue that was brought to light was men accessing the inside of the bulk feed bins. The poor design of the bins meant they were prone to bridging. Banging on the side of the bins was not always effective, so it was necessary to go inside the bins. There was a ladder on the outside of each bin, a lid on the top of the bin allowed access. There was also a ladder inside the bin. It had always been a one man job. The decision was made that it had to be a two man job and if someone was going in a bin it had to be logged at the main farm office. The time, and who was going in the bin and the time out and who was assisting. This was fair enough. The spanner in the works was that the man going into the bin was told he had to have a rope tied around his waist while the man at the top of the ladder outside the bin held on to the rope. The theory was that if the man in the bin passed out or had an accident and could not climb out, the man at the top would pull him out. So, picture this, one man trying to pull up the dead weight of let's say seventy five kilos.

My suggestion was that the home made steel bins should be replaced with modern fibreglass bins that do not bridge.

The method of replacing fuses with nails cropped up too. A certain Volvo driver was identified as the culprit. The company introduced PAT from that point. Something that would be carried out by our very capable in-house electrician.

Several other things were addressed, mainly to do with the use of and storage of medicines and the use of and disposal of hypodermic needles.

It was confirmed that the accidents would be reported to the main office as and when and recorded in one book by the farm secretary. The ensuing health and safety meetings of management proved amusing at times. The reported accidents from the previous month on each unit were read out and discussed. The main reports were of minor needle stick injuries, as in when you manage to self-inject with a hypodermic needle. But one name cropped up regularly. He worked on one of the outdoor units. The funniest entry in the book concerned him. Basically, in his own words,

"A sow stood on my ear when I was weaning." How this could happen baffled everyone. Apparently, he had clambered into a farrowing hut to grab a piglet and the mother had followed him in, knocked him over and proceeded to stand on his ear.

This lads name appeared on a regular basis for various reasons. I guess he was just accident prone.

As it happens the same lad ran my wife off the road in our car, when he was bringing straw in one harvest.

The business manager had some really forward thinking ideas I must admit. He was trying to push for a whole new system to deal with the pig slurry. He found out that there was a group of farms in France that worked under one 'umbrella' as a cooperative. They had this particular slurry system in place that he was interested in.

He very kindly organised a trip over to see how it all worked. Myself, and four of my team from the unit, plus our line manager and the brother who looked after the building maintenance side of the business joined us. We flew from Stansted with Ryanair. It was a time when Ryanair seemed not to use major airports. We could have ended up anywhere. As it happens, we did land at a very small airport which had an 'arrivals hut' rather than an 'arrivals lounge'.

A couple of chaps from the feed company that was heavily involved with the cooperative picked us up in a minibus. They had arranged for us to have lunch at a small restaurant in a village on the way to the main farm. There was a typical spread of French cuisine plus wine. One of the many delicacies was crab claws.

118

There were crab claw pliers provided with them. Nigel picked up a claw and some pliers.

"Monsieur, I wish you to be tres doux, *gentle*, when you are using, please," said our guide.

With that Nigel clamped a giant claw in the teeth of the pliers and squeezed as gently as he could. The shell refused to give, so he applied a little more pressure. Crunch......crab meat and crab shell flew everywhere. I was sat opposite and was splattered. It was hilarious.

Once lunch was over, we set off to see the first farm. All of the buildings had very shallow slurry channels not like the traditional pits that were up to five or six feet deep. Each channel had a scraper on a chain and pulley system that was run every other day. The slurry collected in a conventional holding tank then taken away in tankers to the main treatment plant that was owned by the cooperative. All the farms in the area that were under the 'umbrella' took their slurry to this one plant.

The next stage was amazing. All the slurry was passed through a series of rollers. When the solids came off the end of the process it did not smell at all and was then bagged up and sold as garden compost. The liquid that had been squeezed was almost colourless and was able to be pumped away into the water ways, with no risk of pollution apparently.

This was what our business manager wanted to install on our unit. The capital outlay would have been massive though.

We stayed overnight in a Holiday Inn. The following day a trip round a large abattoir had been arranged. This was also owned by the cooperative. None of us had ever been around an abattoir and it was quite an eye opener. In England the killing out percentage of a pig historically was around 75%. This is what is used for consumption. In France it must have been more like 90%. They use just about every part of the pig, and it was all processed in this one building. I am not going to go into any detail about the various machines in the factory but some of them would not look out of place in a horror movie. I will leave it to your imagination but suffice to say at the end of the production line in the cool storage area we saw a range of things, jars of potted brain, potted tongue, pig's ears and trotters. Hardly anything went to waste. A few delicate stomachs were turned on that tour.

We flew home later that day. It had been a very enlightening trip.

The farmhouse was the second biggest one we had lived in. It was perfect for when all the family visited, but when they had gone there was just the two of us. We had some great fun when they all did come. We bought a big inflatable pool which they had hours of fun in. For one party, which was for one of our grandsons christening, I put a long length of plastic sheet down on the lawn and sprayed it with water and fairy liquid. It made a fantastic slide. I had set up my hi fi outside in the garden and the older grandchildren put on a show. I have been told to mention the 'Sand in my pants' song and dance they did. I have also had a request to mention Edward, one of our twin grandsons. He was five at this point. We had some very large Leylandii trees on the edge of the garden. I think lunch must have been ready.

"Has anyone seen Edward?" his mum asked.

"I've not seen him since we were playing cricket."

"Edward, your lunch is ready where are you?"

After a minute or two a head appeared right at the very top of the tallest Leylandii. I've no idea how he managed to get up there. The branches were so thick and tightly packed.

I thought I had finished book one, but I saw our eldest Granddaughter this weekend and she asked if I had included our visits to Banham Zoo with her and also the fishing, so I have nipped back in from book two. The Zoo was her favourite and we would go there every time her and her mum visited us.

The fishing I have to mention was in a large lagoon on the farm. It was used for irrigating the crops. It had some massive Carp in it. I used to take her down there. We never caught a Carp, but she had a knack of catching the little tiddlers when I was having no luck at all.

There you are Natalie, both mentioned.

On one occasion we came back from a day out and found a lot of pigs in the garden and roaming around outside the workshop that was near our house. The grandchildren thought it was great fun rounding them up. They had broken out of one of the grower pens. The next time the grandchildren came to stay they asked if I would let some pigs out again. I had to explain that they had not been let out last time, they had escaped.

I have just remembered that there was an incident with a French Partridge. My wife had been to work in our Citroen Picasso. We used to park this one in the single  garage with the up and over door. That's what she had done. She came into the house and told me she thought she had hit a pheasant or something. I went outside to inspect the car to see if there was any damage. I opened the up and over door and a rather bedraggled looking French Partridge ran out and took off and flew away into the distance. I looked at the front grille on the car. There were feathers stuck in the grille. The poor bird must have been stuck there all the way up the farm track. Fortunately, without causing any damage.

While we were down there, I decided to sell my Triumph Spitfire. It was a stunning looking car in Lotus yellow with a matt black roll bar. The original metal bonnet had been replaced with a fibreglass Triumph GT6 bonnet. I advertised it in the local paper and within a day or two an American lady rang to say she was interested. I gave her our address and within half an hour she had arrived at our house. I had the car in the single garage. The door was shut when she arrived. I had the roof down on the car so it was looking at its best. I opened the up and over door and before it was even half open, she had made her mind up as she crouched down and caught her first glimpse of it.

"I want it," she said in a very excited state.

"Would you like to take it for a spin?"

"No, it's ok, I'll go home and get the money now."

"You must want to see how it drives."

"Oh, go on then, but you drive,"

We got in and went two miles up the road and turned around in a pub car park and headed back.

"Now I will go home and get the money."

There was no haggling. It was love at first sight. An hour later she arrived back with her husband, handed over the money and away she went. One happy American lady.

It took about a month before I found I had 'classic car withdrawal symptoms'. I decided I really needed a little car in my life. I found a rather sad looking racing green TR7. Not my best ever purchase. I found plenty of rust underneath. I decided to bodge the repairs with plastic padding and fibreglass. I then painted over my repairs with Tetrosyl. It was due an MOT, so I took it to the local garage. I called in later that day to see how it had got on. The mechanic had it raised up on the car lift. He looked at me shaking his head.

"Some idiot has tried covering structural defects with plastic padding," the mechanic said, as he poked a screwdriver through my handy work.

"Who on earth would do something as stupid as that?" I asked, trying to hide my guilt.

"I've no idea, but it's a fail. There is a lot of welding needed if you want to put it through."

I decided to bite the bullet and put it through its MOT. I did not keep the car long after that and sold it for a small profit.

I had my first taste of 'wacky baccy' while we lived down in Suffolk. Gavin had a mate who had a shop where he sold all the paraphernalia for the use of marijuana and other stuff but obviously could not sell drugs. We were chatting about it during one lunch break. He said he would bring a sample for me to try. True to his word, a couple of days later he presented me with one spliff.

I saved it for the weekend. It was a Sunday afternoon, I think. My wife and I were sat out in the conservatory. We had some director chairs in there. We decided to try the spliff. I lit it and took a drag, inhaling deeply. Not a lot happened. I passed it over to my wife and she took a small drag but handed it back to me, she didn't like the taste. I kept puffing away on it until there was nothing left. (big mistake). Gradually it started taking effect. At first, I felt as though my arms were melting into the wooden arms of the director chair. It wasn't a pleasant feeling. It then felt as though my fingertips were bleeding. I needed to go and lay down on our bed. I was totally 'zonked'. I lay there totally helpless. I then felt as though I was peeing myself. I wasn't, but it felt like it. I don't know how long I had been up there but the next thing I knew was my wife telling me my tea was ready and to come down now. I couldn't even move. There was little sympathy forthcoming and needless to say she ate alone that evening.

On the Monday back at work, I told my 'supplier' what had happened.

"Oh my god Mike, you are not supposed to smoke the whole spliff yourself. It was extra strong and had some resin in the tip too".

I've never touched the stuff since and do not want to. We did smoke back then, but not a great amount. We gave up a few times too. On one holiday in Egypt, we had been given up for a good few months. However, we got friendly with a couple of people who smoked. Cigarettes were cheap out there and we ended up starting again. I would imagine plenty of people have done that. Eventually we gave up completely, in 2006 if i recall correctly

Suffolk is a lovely place to live. Summer seemed to last forever again. Winters were almost non-existent. We were only an hour from the coast. There is something about being by the sea that I love. Given the chance I would live down there again.

Four happy years were coming to an end. I was beginning to feel I had achieved everything I could on the farm. It felt like it was time to move on again. As it happens the business was about to start making some serious cut backs because pig prices were plummeting.

And so, yet another move was on the horizon.

This time we would be going much further South.

Current music trend…...The Killers and Black-eyed peas

Fashion…...Shabby chic, miniskirts are back

123

The first offer of a management role came from a farm about as far south as you can get.

We had been looking for a special holiday to celebrate our 25th wedding anniversary. At the same time, I was browsing a magazine that advertised vacancies all over the country and further afield.

The holiday we chose was a two week stay in Sri Lanka. It was all booked up and paid for. We would be going in the February.

It was just before Christmas when I received an email from a farm on North Island New Zealand. I was interested but when the owner rang me, I had to explain that we had a holiday booked and would not be able to afford to fly out there as well.

The owner of the farm said they would cover the loss of the Sri Lanka holiday and pay for both of us to fly out to New Zealand for an interview and to have a look around the farm. This was an offer we decided we could not refuse. So, we cancelled Sri Lanka and booked return flights to New Zealand. We chose to stay for two weeks. I worked out an itinerary, booked our accommodation for our various trips around the island and booked car hire.

The first three days were to be spent on the farm in a house they had empty.

December and January seemed to drag on as our excitement built up. February finally arrived and we finished packing our cases and took the train down to Heathrow. I love flying at the best of times, but one sight of the Jumbo jet that would be taking us on the first leg of our journey, sent excitement levels through the roof. We took off on time and headed for Kuala Lumpur where we would have three hours free before the next leg to Auckland on the North Island. It was a long flight in total. Over 24 hours door to door.

Customs in Auckland airport is rigorous as you probably know. We got through eventually and headed to our pre-booked taxi that took us to the hotel I had booked for the first night.

The following day we had a tour of Auckland which included going up the Sky Tower. That was an experience in itself. The lift that takes you up to the first floor has glass doors. At the

124

bottom you are in a shaft. Then all of a sudden it comes out of the shaft, and you can see everything. I am not good with heights and there was a young boy banging on the doors of the lift. I stood as far back in the lift as I could. On the first floor there was a cafe, my wife decided to wait there while I went up onto the next level. Here you could walk around the circumference of the tower. The walkway over hung the central part and had thick glass panels in it so if you so desired you could stand on it and look down the 300 metres. I kept hold of the grab rail on the inside of the walkway. There was a child jumping up and down on one of these panels, possibly the same one that had been in the lift. There was the option to go up one more level but I declined the offer. A brave bunch of thrill seekers were actually zip wiring down from there. Crazy!

We had another night in the hotel in Auckland then drove down to where the farm was the following morning. Once we were out of Auckland the scenery was amazing. The small towns we drove through all had 'low rise' buildings. We stopped in one town to buy some lunch. It was Otorhanga, near to where the farm was. The buildings were something like you would see in a Western film. Timber clad with porches covering a walkway. The road was wide enough to fit six cars abreast. The impression was of space. Nothing seemed to be crammed in tight. We went into one of the shops and were greeted as though we had lived there all our lives.

We arrived at the farm and were greeted by the farm owner, Bindi, and his business partner, Torvin. I had dressed for interview, shirt, tie, jacket and trousers.

"Welcome to you both," said Bindi, "I'll show you your digs for tonight and for god's sake get rid of the shirt and tie, have you got some shorts? We don't stand on ceremony here Mike."

He showed us to the shack that we would be spending the next couple of nights in. I changed into my shorts and 'T' shirt and made my way back up to the farm office while my wife was looked after by Torvin,s wife. I remember it was a very short interview. Bindi had been in touch with the referees on my CV and had already made up his mind that he wanted me to join his team. We discussed how we could emigrate easily as he would sponsor us as an essential worker. He would help us with housing and provide a vehicle.

The pig unit was well ahead of its time. The feeding was all mechanised. The recording was all computerised. Even the animal welfare side of things was ahead of the UK. There were signs around the unit reminding the staff that, 'when pigs were being moved you use a board, no sticks and no kicking, you can slap them with the palm of your hand'.

Some new buildings were being erected. They were using sectional concrete panels. This was 2004. In the UK concrete blocks were still being used.

After I had been shown around the pig unit Bindi asked Torvin to show my wife and I the area to get an idea where we might end up living. Everywhere looked so fresh and clean. All the properties seemed to be bungalows set on large plots of land. Torvin told me I would be given a pick-up truck like the one he had. It all seemed too good to be true.

When we got back to the farm office all three of us sat down with Bindi.

"So, Mike, the job is yours if you want it."

My brain was now in overdrive.

"Obviously we do not expect you to give us your answer right now. Go and enjoy the rest of your holiday and enjoy your wedding anniversary. Get back in touch with us when you get home. It is a major move we realise."

We both left the office in a bit of a daze. We packed our cases and I loaded the car up. Before we left we said our thank yous and goodbyes. As we drove away my wife turned to me and said,

"What just happened!?"

"I think I have just been offered a job 12000 miles from home."

"Yes, I thought so."

We had a lot to consider. New Zealand is an amazing place. It was the opportunity of a lifetime. We decided to enjoy the rest of our holiday before making any decisions. There was so much to see but we would barely scratch the surface by only seeing a small section of the North Island.

126

Our first port of call was a hotel near Waitomo for our first night away from the farm. On the way we decided to take a detour to a place called Kawhia. It is on the West coast and sits in a large bay. We pulled up in a small car park that I reckoned was near the bay. The car park was dwarfed by massive sand dunes that must have been thirty feet high. It was really soft sand, and my wife decided she would not manage the climb. I simply had to. So off I went. I was like a child on his first trip to a beach. The climb was worth it. I reached the top and my heart skipped several beats. The view over the Tasman Sea was breathtaking. The vast expanse of golden sand stretched for miles and miles in both directions and was deserted. The tide was out and must have been one hundred metres away. A paddle in the Tasman Sea was obligatory.

We spent that night in a lovely room in the turret of the hotel.

For our second night I had booked a lodge on the northeast shore of Lake Taupo. Taupo is on the edge of the Tongariro national park. It is a volcanic region with thermal springs bubbling up around the edges of the lake. In the lodge there were leaflets illustrating things of interest in the area. Several caught my eye. The first we were tempted to try was a light aircraft flight over the national park and lake. We had never been in a small aircraft.

I made a phone call and booked the trip for that afternoon. We arrived at the grass airstrip and were greeted by a young man who looked like he may have just left school.

"Hi, you must be Mike and Mrs Mike, follow me and let's get airborne,"

We signed an insurance form and headed to the small plane that sat on the edge of the airfield. I was looking around for the pilot. It was the young man. The plane was a four seat Cessna Skylane. My wife sat in the front with the pilot, and I sat behind her. We put on our earphones and taxied ready to take off. It was a pretty bumpy grass strip and as we lifted up into the clear blue sky we were buffeted.

"Is it always this turbulent?" I asked. Our young man laughed.

"Oh, this isn't turbulence, wait until we get over the volcano."

He took us Southwest from the lake and circled around the national park and over Mt Ruapehu, one of the active volcanoes.

It was jaw dropping. He then took us around the circumference of the lake then over Taupo and back to the airstrip. My wife often reminds me of how much she enjoyed it and the fact that I was holding on with white knuckles. This was one of the best one hundred and fifty pounds we have ever spent.

While we were in Taupo we went to the Geothermal baths. There were several small baths and one large one supposedly with natural healing powers. They certainly were very relaxing.

Another site that was well worth a visit was 'Craters of the Moon'. The whole area had lots of thermal activity with geysers bubbling up and often shooting large amounts of liquid clay up into the air.

My favourite visit was to The Huka Falls. These are on the Waikato River. I took a jet boat ride which takes you to the foot of the falls. We were told that the force of the water crashing down caused the water to be so full of air that the boat would not stay afloat if we went too close.

After two days in Taupo, we decided to drive to the East coast. I had read that the town of Napier was well worth a visit because of its Art Deco style buildings. We were slightly disappointed when we arrived. The town looked very run down unfortunately.

We didn't stay long and set off up the Main highway heading for Rotorua and then on to Tauranga. This was quite a long drive. I managed to shoot three hours of film on my camcorder while my wife drove. I was so captivated by the scenery. We drove up to Coromandel where we spent our penultimate last two nights at the Jacaranda Lodge.

Our holiday was coming to an end. We drove over to Auckland and had one last night at the same hotel we had spent our first night. The following morning I handed over the hire car and took a taxi back to the airport.

It had been an amazing two weeks. We took the train from Heathrow to Kings Cross via the underground then boarded our train to take us back to Stowmarket. In an attempt to bring us weary travellers back into the real world, British Rail had decided to undertake essential maintenance work on the line between Ipswich and Stowmarket. We had to complete the final leg of our epic adventure on a coach!

While we were away we did discuss how such a big move would fit in with leaving family in the UK. We pondered over whether my wife could visit home a couple of times a year. Once we were home reality set in and leaving parents, children and grandchildren seemed to be not such a good idea after all. We had a few sleepless nights trying to weigh up the pros and cons. We finally came to the really difficult decision that I would have to make the call to New Zealand to decline the very generous offer I had been given. If the opportunity had arisen when our children had been younger we proberbly would have gone.

In the meantime I had been putting feelers out amongst people I knew in the industry to see if there were any vacancies around.

A feed representative who I had known for many years contacted me about a management position he had heard about. He had contacts with a large firm based in Segovia in central Spain. The company had feed mills, an abattoir and several pig units spread around central Spain.

Apparently, the company wanted to get some 'English blood' to manage their largest unit. It was a 900 sow Nucleus unit. This type of unit is at the top of the 'health pyramid' and supplies breeding gilts to multiplier units. A multiplier unit is what the unit at Crayke was.

I expressed an interest and was invited over to Segovia for an interview. I booked my flight. Stansted to Madrid, where I would pick up a hire car. I had been given detailed directions and I printed them off to avoid getting lost. When I finally negotiated my way out of the Airport I set off on the motorway.

I had been told to keep on the AP6 then take the AP61, after going through the tunnel under the Guadalajara mountains.

I seemed to have been travelling for far too long and panic set in. I took the exit onto the M601. Big mistake. It turned out to be a very long way to Segovia over the mountains. I had to pull the car over and ring the company's office to explain that I was lost and was going to be late. I was told to just carry on that road but I had just added two hours onto my journey. This was not a good start.

Fortunately, the interview went better than my map reading. Mateo who was the head of the scientific research department of the company was instrumental in wanting to bring over some new

129

blood into the company. He was a really friendly chap. After the interview we called in at a tapas bar for a beer and some tapas. This was my first taste of 'real Spain'.

I stayed overnight in a small hotel and the following day was shown around a couple of the pig units that were underneath the company's umbrella.

This was a real eye opener. I must say I was not that impressed. There seemed to be a lack of animal welfare. Ironically, I was being shown around by one of the company's vets who seemed to think everything looked ok. I will come back to animal welfare Spanish style later.

At the end of my two day visit I was offered the job. Two job offers in a space of two weeks. I was feeling pretty chuffed with myself. This offer would entail a more realistic move with it only being a two and a half hour flight away from family. However, there was still some serious thinking to do.

When it was time to set off back to the airport Mateo was concerned that I may get lost again, so he suggested I follow him to the motorway having told me to stay on this one and take junction 12 which would lead me straight to the airport. He waved me off and turned around to go back to Segovia. I had allowed myself plenty of time.

I was tootling along watching out for junction 12 having negotiated the tunnel this time. I knew I was getting close. Junction 9, ... Junction 10, ... Junction 11, ...Junction 13.... What? Panic again ensued. I was now heading into central Madrid having come off the motorway at Junction 14. I was lost. I pulled over to ask directions from a couple of chaps sitting on a bench on the side of the road. They had no idea how I could get to the 'Aeropuerto'. Time was now not on my side. I even considered abandoning the hire car and taking a taxi from where I was. I decided against that and drove on. I was now in a very built up area. Just as I was about to tackle a large roundabout, I spotted a young couple walking on the pavement. I stopped and wound my window down and asked in my terrible Spanish how to get to the airport. They spoke almost perfect English and pointed out that I was only a couple of miles away. Just off the first exit of the roundabout was a large, very welcome, sign. AEROPUERTO. Thank God. I eventually found the car hire office and handed over the keys. And relaxed.

When I got home, we mulled it all over and weighed up the pros and cons once again. The financial side of the offer was very generous. There would be a house provided free of charge and a vehicle. We decided to go ahead with a view to making this a long-term position. We could live relatively cheaply over there and bank lots of money in the process.

My wife and I were invited over to have a look at houses. It was really cold while we were there. -17 deg C. The first village we looked at was a fairly new build, but it was deserted. We decided it was too quiet! This was our first mistake. We ended up choosing a house in a very small village where it turned out we would be the only English residents in a community of 650 people.

More details of life in Spain will follow later.

So, the decision was made. I handed in my notice and started planning for our biggest move ever. Downsizing was essential. Fortunately, there was a regular car boot sale in a village near where we lived. I managed to sell lots of our 'surplus to requirement belongings'. It's amazing what people will buy at those events. Some of the people who attend are serial buyers. As soon as you open your boot they swarm around you like Vultures.

One funny sale I made was a large glass carboy. I had it priced at ten pounds to start with. There was some interest but no sale. I changed the price tag to five pounds. That did not work. I didn't want to take it back home, so I made a new sign saying 'free to a good home'. I had just about given up when a chap came along and said he wanted it.

"I saw you had five pounds on it earlier."

"Yes, that's right, but I have reduced it."

"I'll give you five."

"Ah, ok."

That's Suffolk for you.

And so, after several car boot sales and the odd garage sale we were left with what we needed for our move. The last thing to go was our Citroen Picasso which we sold to my brother.

131

As our moving date grew close, we had most things organised. I had booked a Luton box van and the ferry crossing from Plymouth to Santander.

I decided I needed to thank everyone associated with the business I had been working with for the last four years. My wife and I invited everyone around to our house for drinks and snacks. I think they were all genuinely sorry to see me moving on. It had been a truly happy four years and we still think of Suffolk as our second home.

Dan and his wife very kindly offered to help with our move. My wife went to stay with our daughter while we tackled the process of packing all our belongings into the van. In order to make sure everything would fit in the van I had made scale cut out shapes of all the boxes, furniture, beds etc.

It all fitted in perfectly, albeit very snug. Every cubic inch was taken up. The last thing to go in was the vacuum cleaner. There was only just enough room for it but I managed to hold it in place as Dan pulled down the roller shutter door.

That was it. All loaded and time to hit the road.

"I have a good padlock we can use Mike."

"Brilliant Dan, I hadn't even thought about that."

Dan locked up and kept hold of the key! We picked Dan's wife up and set off for yet another epic adventure. An adventure that was to prove to be more of a challenge and more stressful than my wife and I could ever have imagined..........

11588 days have passed....... To be continued......

Glossary of terms......

Gilt.... Female pig that has not yet had a litter. Although often still called a gilt until weaned and served again.

Sow...Female pig that has had one or more litters

Boar...Male pig of any age.

Service...The act of insemination, either artificial or natural.

Farrowing ... The act of the female giving birth.

Farrowing house...Maternity ward

Dry sows...A term used for pregnant sows

Weaning...The act of taking piglets off the mother. Normally around 25 to 28 days of age

On heat...When the female is ovulating. This occurs every three weeks unless insemination was successful. A sow will only serve when she is on heat.

Flat decks / Weaner house...Accommodation for weaned piglets. Housed up to around 35kg

Fattening house...Accommodation for growing pigs from 35kg to pork size 70kg or bacon size 95 kg

Spring tine harrow...An implement used on the back of a tractor to break the soil down ready for sowing seeds. Now mainly outdated as machinery has advanced.

Ad lib feeding...Pigs have access to food and water 24/7

Printed by BoD™in Norderstedt, Germany